Letts

Revise KS3

Science

Bob McDuell and Graham Booth

Contents

Science at Key Stage 3

Introduction to KS3 Science

This Science Study Guide has been written specifically to provide complete coverage of the Science National Curriculum at Key Stage 3. The content has been carefully matched to the QCA Scheme of Work from Year 7 up to and including Year 9.

This is an important stage in your education, because it lays the foundation for the Science that you will need at Key Stage 4, when you are studying for your GCSEs. The National Curriculum requires all 14 year olds to follow the same programme of study, which defines the knowledge and skills that you need to learn and develop.

Science is split into four Attainment Tasks:

Sc1 Experimental and Investigative Science

You will carry out practical activities in school concerned with planning, carrying out, analysing and evaluating experimental procedures. Some of these activities will include ICT. Some of the questions in tests will also assess parts of Sc1.

Sc2 Life Processes and Living Things

This area of Science is called Biology. You are required to study how the human body works, how plants work, the environment, how living things are grouped and how species change from generation to generation.

Sc3 Materials and Their Properties

This area of Science is called Chemistry. It includes classifying materials in different ways, changing materials (including chemical reactions) and grouping materials according to their properties.

Sc4 Physical Processes

This area of Science is called Physics. It includes electricity and magnetism, forces and motion, light and sound, energy resources and energy transfer, and a study of the Solar System.

At the end of Key Stage 3, you will take tests that are set for all students in England and Wales. Your performance in these tests will be awarded an overall level. Most students at the end of Key Stage 3 will achieve a level of 5 or 6. For more exceptional performance, levels 7 and 7★ can be awarded.

The results may be used by the government to check whether standards of achievement are rising nationally and by people who make comparisons of the achievements of pupils in different schools.

Beyond Key Stage 3, you will start GCSE courses in Double Award Science, Single Award Science or separate sciences.

How this book will help

Successful study

This book should be used to help you throughout Key Stage 3 to make sure you know and understand the key facts and issues as you go along. You should then be able to use this information to help you answer questions.

Success in school depends on regular planned work over a period of time rather than panic bursts of very hard work just before an examination.

If you develop the habit of regularly reviewing the work you have done in school, and making sure you understand it, it is less likely to cause you severe problems as the SATs approach.

Features in the Study Guide

Chapters 1, 4 and 7 cover the work you are likely to study in Year 7.
Chapters 2, 5 and 8 cover the work you are likely to study in Year 8.
Chapters 3, 6 and 9 cover the work you are likely to study in Year 9.

Progress checks

Throughout the book there are Progress Check questions. These are not of the type that you will meet in the SATs but are there to help you check what you have learned as you go through the topics in the book.

Key points

There are Key Point panels in each topic. These panels draw your attention to important facts and definitions.

Margin comments

Margin Comments contain invaluable advice and guidance from the examiners.

Practice questions

At the end of each chapter there are a couple of practice test questions to give you a clear idea of the types of questions you will be asked.

Preparing for Standard Assessment Tests (SATs)

Planning your time before the exam

When you walk into the examination room to take your Science tests, you want to feel confident in your own ability to answer the questions. You will not feel confident if your 'revision' has been limited to the evening before the test.

How to revise Science

At the beginning of the Easter holidays, draw up a revision plan. Start with the Science subject that you understand least well, so that you have time to revisit it. There are five major topics in Biology, three in Chemistry and five in Physics. As these topics vary in length, it is best to allocate a week for revising each of these areas.

- Using this Study Guide, list, read through and make **brief** notes on the Biology topics in Years 7 to 9. Keep the extent of these notes down to approximately one side of A4 paper for each year. Underline and highlight the main points in the notes as necessary.
- Do the same for the Chemistry and Physics topics.
- Now the hard part – condense your notes to just one side of A4 paper for each of Biology, Chemistry and Physics.

Different types of questions

The different types of questions that you will have to answer include:

- Questions that test your knowledge – you may be asked to identify the labelled parts of an organ system, such as the respiratory system.
- Questions that test your understanding – you may be asked to explain how the seasons are related to the Earth's tilt.
- Questions that test your awareness of safety – you may be asked about the safety precautions necessary when carrying out an experiment.
- Calculations – you should be able to calculate the size of physical quantities, such as speed and pressure.

Getting ready for the exams

You don't want to be flustered on the morning of your exams, so make sure that you have everything ready the night before:

- two sharp pencils and two pens
- a rubber and a ruler
- a protractor

Do a final check by writing out 10 questions – the things that you find most difficult to remember. Test yourself two or three times on these. Then, have a good night's sleep.

GOOD LUCK!

Biology

Chapter One (Year 7)		Studied	Revised	Practice Questions
1.1 Cells	What are cells? Cells, tissue and organs Specialised cells			
1.2 Reproduction	Adolescence Menstruation Fertilisation The developing fetus			
1.3 Environment and feeding relationships	Habitats How plants and animals are adapted Feeding Changes in the environment			
1.4 Variation	Different organisms Nature or nurture			
Chapter Two (Year 8)				
2.1 Food and digestion	A balanced diet Vitamins Mineral salts Testing for proteins, carbohydrates and fats The digestive system			
2.2 Respiration	Respiration in human cells What is the role of the lungs? Inhaled and exhaled air			
2.3 Microbes and disease	How the body prevents micro-organisms entering			
2.4 Classification	Classification of animals Classification of plants			
Chapter Three (Year 9)				
3.1 Inheritance and selection	Passing on information to the next generation Variation Mutations Selective breeding Asexual reproduction			
3.2 Fit and healthy	Fitness Breathing Effects of smoking Effects of alcohol Effects of drugs The working of simple joints			
3.3 Plants and photosynthesis	Photosynthesis			
3.4 Plants for food	Minerals needed for plant growth Predator–prey relationships Pesticides Advantages and disadvantages of growing in a greenhouse			

Year 7

1 Chapter One

The topics covered in this chapter are:

- **Cells**
- **Reproduction**
- **The environment and feeding relationships**
- **Variation**

1.1 Cells

After studying this topic you should be able to:

- **describe the structure of animal and plant cells**
- **explain the job of each part of a cell**
- **describe how cells form the structure of animals and plants**
- **explain how some cells are adapted to carry out their job**

What are cells?

Just as a house is built out of bricks, animals and plants are made up of **cells**. Cells are the building blocks of all living things.
There are differences between the cells that make up animals and those of plants. The diagram shows typical plant and animal cells.

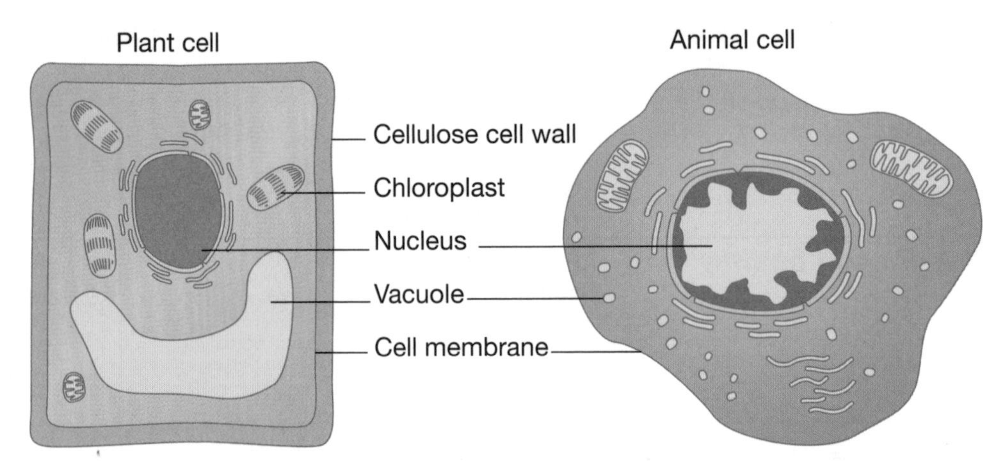

Fig. 1.1 Plant and animal cells.

Animals have a **skeleton** to keep them in shape. Their cells are soft to the touch. Plants do not have a skeleton – they stay upright because they have a tough wall around each cell.

Key Point

The main difference between plant cells and animal cells is that plant cells have a **cell wall**.

The exception is red blood cells, which do not have a nucleus.

All cells have:

- a **nucleus** that controls the actions of the cell
- **cytoplasm** – a mixture of chemicals that includes nutrition that has passed into the cell and waste products on their way out of the cell
- a **cell membrane** – this allows nutrients in and waste products out

In addition, plant cells have:

- **chloroplasts** – these contain a green chemical called **chlorophyll** that absorbs energy from the Sun. This energy is needed for the plant to make food
- a **cell wall** to keep the cell in shape

Progress Check

1 Decide whether each statement is true or false.

a Animal cells are rigid, but plant cells are floppy.
b Only plant cells have chloroplasts.
c The nucleus of a cell controls its actions.

2 What is the main difference between plant cells and animal cells?

3 How do waste products leave an animal cell?

1. a False b True c True 2. Plant cells have a cell wall.
3. By diffusion through the cell membrane.

Cells, tissue and organs

There are lots of different types of cell that make up an animal or plant.

- Cells of the same type are grouped together to make **tissue**.
- All the cells in a tissue carry out a similar task.
- Several different tissues make up an **organ**.

Key Point

Cells make tissue and tissue makes organs.

Your body is made up of a number of organs, each of which has a different job. Some important organs are:

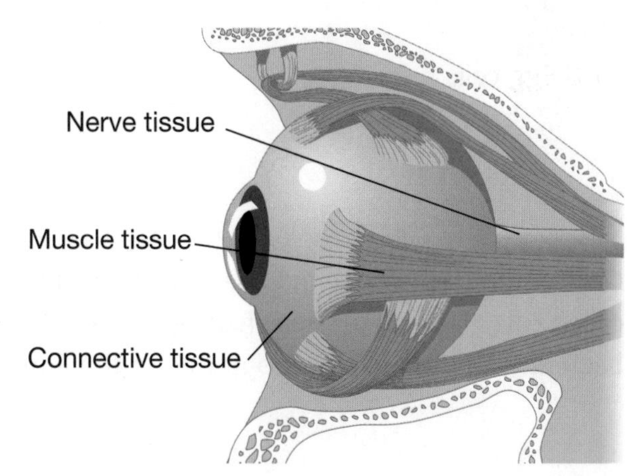

Fig. 1.2 The eye.

- the eye – this contains **muscle tissue** to move the eyeball, **nerve tissue** to send information to the brain and **connective tissue** that forms the 'white' of the eye
- the skin – this contains **nerve tissue**, **sweat gland tissue** and **blood vessel tissue**
- the heart – this contains **lining tissue**, **tendon tissue** and **connective tissue**

The diagram shows some of the tissues in an eye. You can see how organs are made of different types of tissue, each of which has a different job to do within the organ.

Specialised cells

Questions about ways in which cells are adapted to do their job are common in tests and examinations.

Cells in different tissues have different jobs to do. Their shape is adapted so that they can do their job efficiently.

The diagrams show some specialised cells and how they are adapted to do a certain job.

Cells in the underneath part of a leaf have very few chloroplasts.

The job of root hairs is to absorb substances from the soil.

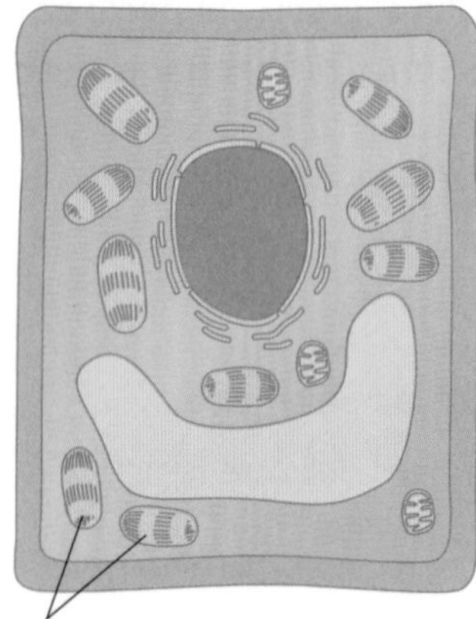

This cell from the **upper surface of a leaf** has lots of chloroplasts to absorb energy from the Sun.

Fig. 1.3 Leaf cell.

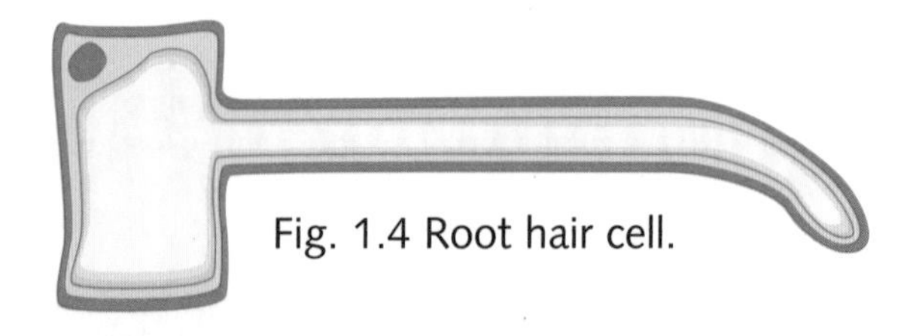

Fig. 1.4 Root hair cell.

This **root hair** cell has no chloroplasts – it is underground. It has a large surface area so that water and minerals can pass into the cell.

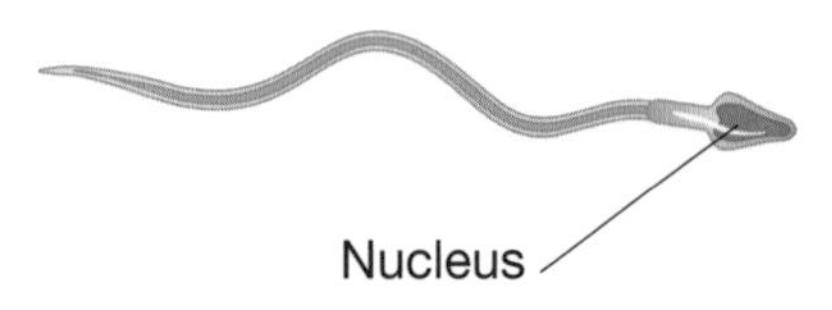

Fig. 1.5 Sperm cell.

A **sperm** cell is a male sex cell. It needs to travel a long way so it has a tail to propel it and a streamlined shape.

A hen's egg is much larger than a human egg because it also contains the food needed for the fertilised egg to grow.

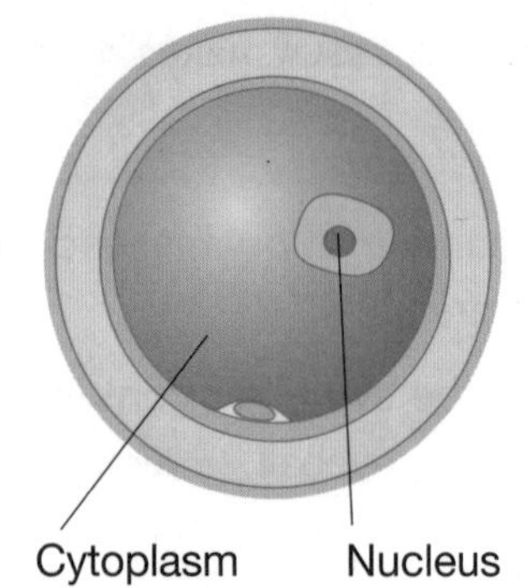

An **egg** cell or **ovum** is a female sex cell. It is much bigger than a sperm. It does not have to propel itself and it only moves slowly.

Fig. 1.6 Egg cell.

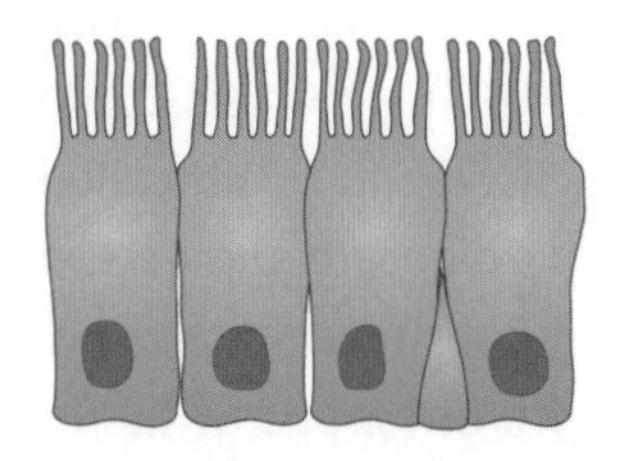

Fig. 1.7 Cells with cilia.

Cells in the **trachea** or **windpipe** have tiny hairs called **cilia**. The job of these hairs is to trap bacteria and dust so they do not enter the lungs.

Progress Check

1 What is the name for a group of similar cells?
2 An organ is made up of different tissues. True or false?
3 What is the job of the chloroplasts in a plant cell?
4 Why do human cells not contain chloroplasts?

1. Tissue 2. True 3. To absorb energy from the Sun. 4. Humans do not make their own food.

1.2 Reproduction

After studying this topic you should be able to:

- **describe the changes that take place during adolescence**
- **explain what happens during the menstrual cycle**
- **explain fertilisation and describe how a fertilised egg develops into a baby**

Adolescence

New humans are created by **sexual reproduction**.

Key Point

In sexual reproduction a male sex cell joins with a female sex cell to create a new living thing.

> Some plants and animals can reproduce without sex. This is called asexual reproduction. An example is growing a new plant by taking a cutting from an existing plant.

> The hair that grows around the penis and the vulva is thicker than the hair that grows on the head. It is called pubic hair.

Before they can reproduce, children need to change into adults. This happens during **adolescence**.

In boys the main changes that take place between the ages of 11 and 14 are:

- They start to produce male sex cells (sperm).
- The voice becomes deeper.
- Hair grows around the **penis** and on the **scrotum**.
- The penis becomes larger.

In girls, adolescence may start earlier, from the age of 10. The main changes that take place are:

- They start to release female sex cells (ova) from the **ovaries**.
- **Breasts** develop.
- Hair grows around the **vulva**, the fleshy external opening to the **vagina**.
- A regular **menstrual cycle** begins, with **periods** each month.

Both boys and girls change emotionally during adolescence. They become sexually attracted to other people and are often concerned about whether they are sexually attractive to others.

Progress Check

1 Complete the sentences:

In adolescence boys produce male sex cells called _______.
In adolescence girls release female sex cells called ______.

2 In women, periods occur once each:

A – day B – week C – month D – year

Which option is correct?

3 What change takes place to a boy's penis during adolescence?

1. Sperm, eggs or ova 2. C 3. It becomes larger.

Menstruation

In diagrams of the female reproductive system, pupils often confuse the cervix with the vulva. The vulva is the external entry to the vagina, the tube that leads to the uterus.

A period is a result of the **uterus**, or womb, preparing itself each month to receive a fertilised egg.

The diagram shows the main parts of the female reproductive system.

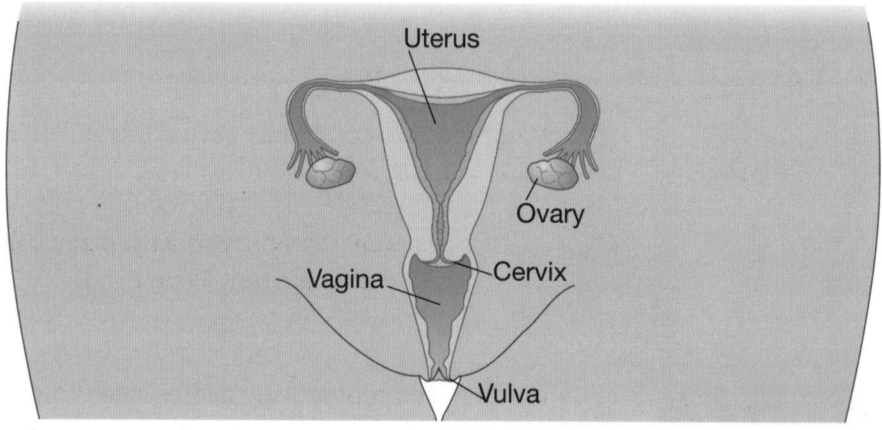

Fig. 1.8 The female reproductive system.

Girls are born with thousands of unripened eggs, but adult males are continually making new sperm.

The lining of the uterus is at its thickest as the egg enters the uterus from a fallopian tube.

In the menstrual cycle:

- An egg ripens and is released from an ovary each month.
- As the egg travels slowly along the fallopian tube, the lining of the uterus thickens.
- If the egg enters the uterus unfertilised the thickened lining falls away.
- The unfertilised egg and the lining of the uterus pass out through the vagina – this is the period.

The diagram shows the changes that take place during the menstrual cycle.

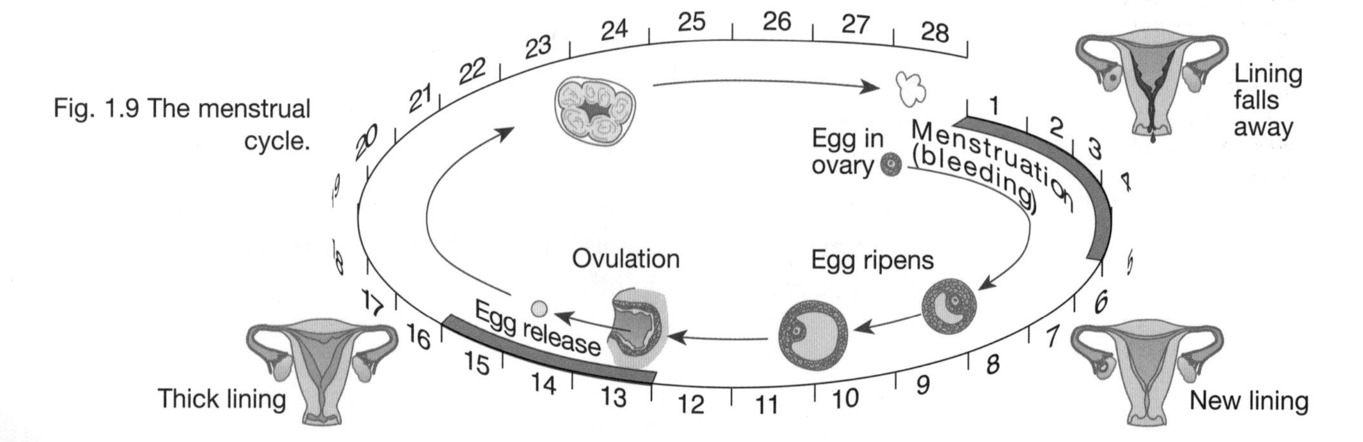

Fig. 1.9 The menstrual cycle.

Progress Check

1 Choose the best word to complete the sentence:

The uterus is also known as the ______________.

bladder **vagina** **vulva** **womb**

2 In the menstrual cycle, the lining of the uterus fills with blood ready to receive an unfertilised egg. True or false?

3 What happens to the thickened lining of the uterus during a period?

4 What is the name of the entrance to the uterus from the vagina?

1. Womb 2. False 3. It passes out through the vagina. 4. The cervix

Fertilisation

When the penis is erect, it is at the correct angle to fit into the vagina.

The diagram shows the main parts of the male reproductive system.

Sperm tube
Erectile tissue
Penis
Testis
Scrotum

Fig. 1.10 The male reproductive system.

During **sexual intercourse**:

- the **erectile tissue** becomes filled with blood, causing the penis to swell and become **erect**
- the penis is inserted into the vagina
- sperm from the **testes** flows through the sperm tube and out through the end of the penis
- the sperm are normally released at the **cervix**, the entrance to the uterus

In plants, male sex cells are called pollen.

The sperm swim through the uterus and along the fallopian tubes. If they meet an egg in the fallopian tubes, fertilisation may take place.

Key Point

Fertilisation is the joining together, or fusion, of a male sex cell (sperm) and a female sex cell (ovum or egg).

When an egg is fertilised, one sperm enters the egg and the nuclei of the two cells join together. This single cell is the beginning of a new human. Its nucleus contains information from both the mother and the father.

The diagram shows what happens in fertilisation.

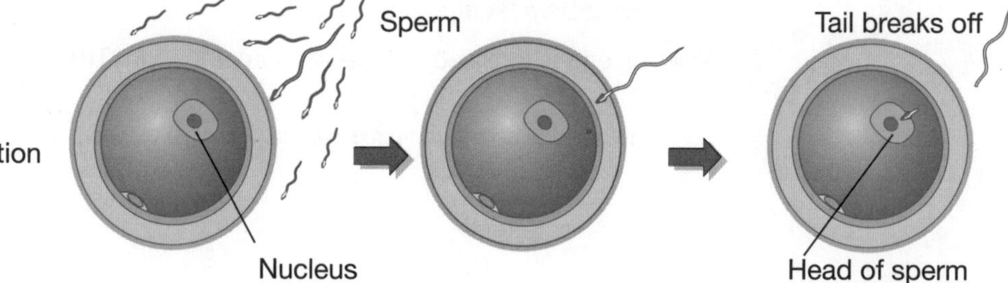

Fig. 1.11 Fertilisation

The single cell divides and these cells divide several times as they travel along the fallopian tube to the uterus. Now it is an **embryo**.

The developing fetus

> The embryo becomes a fetus when it has recognisable human features.

> It is important to remember that the blood of the fetus and that of the mother never mix.

The embryo travels to the uterus and beds itself in the lining. Here the embryo develops into a **fetus**. The fetus:

- is joined to the **placenta** by the **umbilical cord**; food, oxygen and waste materials pass along the cord
- is surrounded by a watery liquid that keeps it warm and protects it from shock

After nine months of **pregnancy**, the baby is born through the vagina. A human baby is helpless at birth and relies on care from parents if it is to survive. Some animals produce babies that do not need a great deal of care from parents.

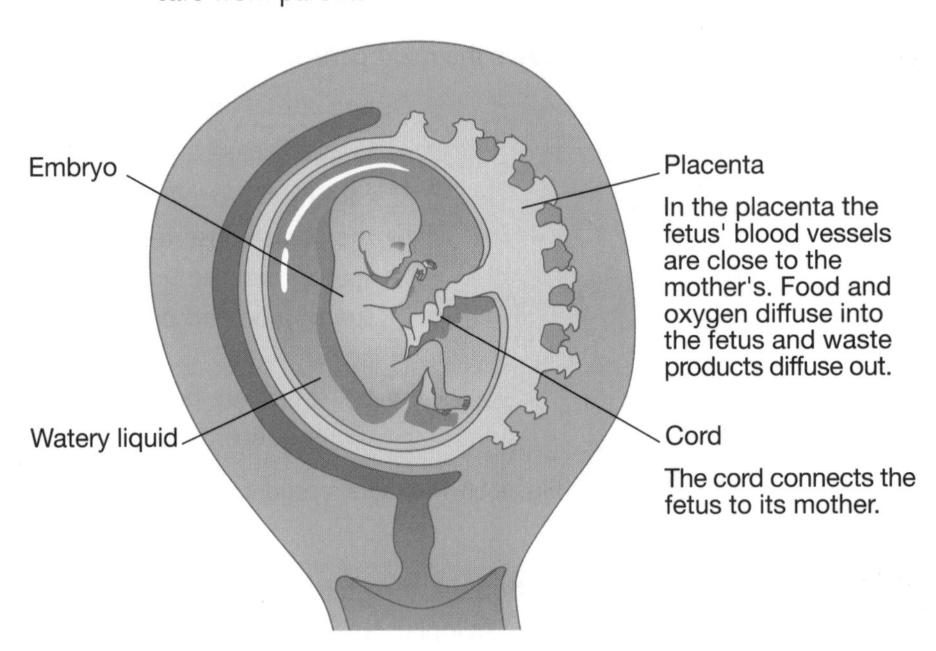

Fig. 1.12 The developing fetus.

Progress Check

1. Which of the male reproductive organs produce sperm?
2. Complete the sentence:
   The testes are contained within a bag called the __________.
3. In the placenta, how does food pass into the blood of the fetus?
4. What connects a fetus to the placenta?
5. When an egg is fertilised, it fuses with another egg. True or false?

1. The testes 2. Scrotum 3. Food from the mother's blood diffuses through the placenta.
4. The umbilical cord 5. False

1.3 Environment and feeding relationships

After studying this topic you should be able to:

- **describe how a habitat provides the environment that animals and plants need to survive**
- **explain how the animals and plants in a habitat depend on each other**
- **describe how some organisms are adapted to survive changes in their habitats**

Habitats

A pile of rotting wood provides all these things for a colony of wood lice.

For much of the year, we live indoors. Most plants and animals live outdoors. The region where a particular plant or animal lives is called its **habitat**. A habitat provides:

- food
- shelter
- a place to reproduce

Key Point

A habitat is a place where an organism lives, feeds and reproduces.

In a large garden there can be a number of habitats:

- A hedge provides food for worms and other soil-based animals that feed on the dead leaves.
- It provides shelter for birds and hedgehogs.
- The birds and hedgehogs eat the animals that feed off the hedge.
- Other habitats could include a tree or a pond, or an area of bushes or flowers.

Progress Check

1 What are the three important things that a habitat provides?

2 Plants are important to a pond habitat because they add oxygen to the water. True or false?

3 How does the fox's habitat provide it with food?

1. Food, shelter and a place to reproduce. 2. True
3. The fox's prey, smaller animals, live in the habitat.

How plants and animals are adapted

You will often come across questions in tests about how different plants and animals are adapted to their habitat.

The animals in any habitat have special features that enable them to live there. A fish can live in a pond but a fox could not. To live in a pond a fish needs:

- fins so that it can move through the water
- a tail so that it can control its direction of movement
- gills for gas exchange

A fox needs:

- strong hind legs so that it can run fast to overtake its prey
- canine teeth to tear flesh when it feeds
- lungs for gas exchange

Bluebells and anemones are woodland plants that flower in spring.

Plants are also adapted with special features that suit their habitat. Woodland plants often develop and flower in the spring when light can penetrate through the trees. In summer, when there is little light, they are dormant.

The plants in a pond grow most vigorously in summer when there is the greatest amount of sunlight, which gives them energy for making food and growth. They are adapted to absorb energy from the Sun by floating on the water or having long stems so that their leaves are close to the water surface.

Feeding

Plants obtain essential vitamins and minerals from the soil.

Food is important to all plants and animals. It provides the energy needed to keep their organs working and to allow movement and growth. It also provides animals with the essential vitamins and minerals needed by different organs in the body.

Key Point

The key difference between plants and animals is that plants make their own food.

Plants use **carbon dioxide** from the air and **water** from the soil to manufacture food in the form of **glucose**, a simple sugar.

- Because they make all the food for the animals in a food chain, plants are called **producers**.
- Animals such as worms and slugs that feed directly off plant material are called **primary consumers**.
- **Secondary consumers** feed off primary consumers.
- **Tertiary consumers** feed off secondary consumers.

As you can see, some animals, called **herbivores**, feed off plants. Other animals, called **carnivores**, feed off other animals. There are also animals, such as humans, that feed off both plants and other animals. These are called **omnivores**.

The feeding relationships in a habitat are shown by a **food web**, which consists of a number of **food chains**.

The diagram shows a food web based on an oak tree as the **producer**.

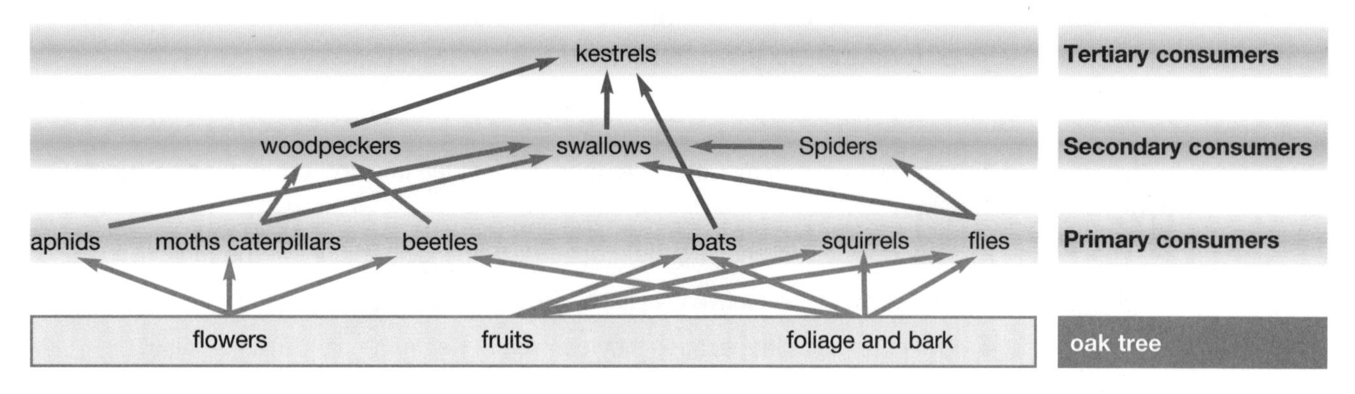

Fig. 1.13 A food web.

A common error in tests is to draw the arrows the wrong way.

One of the food chains that make up this web is:

flower → moth caterpillar → woodpecker → kestrel

Key Point

The arrows on a food chain or food web show the direction of energy flow.

1 Complete the sentence:

The producers in a food web are always ___________.

2 Why is it an advantage for a fish to have a streamlined shape?

3 In a wood, there are more swallows than kestrels. Suggest why.

1. Green plants 2. So that it can move quickly to escape predators
3. Kestrels need a lot of swallows to feed off as they eat several each day.

Changes in the environment

If you observe a habitat such as a hedge or a tree in a garden, you notice that changes are continually taking place.
Each day there are changes in:

- the light intensity
- the temperature
- the moisture level

Over the course of a year there are even greater variations in all these factors. These changes can affect:

- the number of organisms in a habitat
- the type of organisms in a habitat
- the activity of the organisms in a habitat

Evergreen trees often have needle-shaped leaves to minimise the water loss.

To survive in a changing habitat, organisms need to be adapted to cope with the extreme conditions. Many trees lose leaves in winter to reduce the loss of heat and water – it may not be able to take in water if the ground is frozen. Insects burrow into the ground or leaf litter, and their offspring survive the winter as eggs. Large animals such as squirrels build up a store of food to help them to survive.

Progress Check

1. In a cool wet summer there are more snails than in a hot, dry summer. Suggest why.
2. How does burrowing into the ground help an insect to survive the winter?
3. All trees lose their leaves in winter. True or false?
4. Owls hunt at night because they can see better in the dark. True or false?

1. Snails need cool, damp conditions to reproduce. In hot weather they die and do not reproduce.
2. The ground insulates the insect and stops it from freezing. 3. False 4. False

1.4 Variation

After studying this topic you should be able to:

- **describe how individuals of one species, e.g. humans, vary**
- **describe how animals and plants vary to form different species**
- **explain the causes of variation**

Different organisms

Human beings are a **species**, which is a type of organism.

Fig. 1.14 Human beings can differ.

All human beings:

- have a head, two arms and two legs
- walk on two legs
- reproduce sexually
- breathe with their lungs

There are some individual exceptions to this.

The list is endless, but what about the ways in which human beings differ? Here are some obvious ones:

- height
- weight
- colour of hair or skin or eyes

Again, the list is endless.

Key Point

Differences between species and between individuals of a species is called variation.

Nature or nurture

Some people have eyes of different colours – this can be caused by the effect of chemicals in the environment.

If you have some family photographs, you can compare your own physical appearance with that of your parents when they were the same age.

What determines whether you are a human being, rather than a fly or a cat or a rabbit? The answer is the **genetic information** from your parents. This information was contained in the sperm and the egg that joined at the instant you were created. The same information is contained in all the nuclei of your body cells. This information determines:

- the colour of your hair
- your sex – whether you are male or female
- the colour of your eyes

It also influences a number of other things, such as:

- how tall you are
- your body shape
- your weight

Genetics is not the only factor that determines your height, shape and weight. These are also influenced by **environmental factors** such as:

- your diet – whether you eat a balanced diet or a lot of fatty foods and chocolate
- exercise – for example, whether you walk or cycle to school or travel in a car
- physical activity – whether you take an active part in sport or prefer to watch television

There are some aspects of your physical appearance that have nothing to do with genetics – they are totally controlled by environmental factors. These include:

- the length of your hair
- whether you wear an earring
- the length of your finger nails

Most of the differences between human beings are due to a combination of genetic and environmental factors – nature and nurture.

Progress Check

1. A species is:

   A – an animal B – a plant C – a type of organism

   Which is correct, A, B or C?
2. Variation describes the ways in which different organisms are similar. True or false?
3. Which TWO options are correct?

   A genetic information is inherited by an organism from its parents
   B genetic information is controlled by the environment
   C genetic information is contained in every nucleus in body cells.
4. Is the colour of a person's skin controlled by genetic factors, environmental factors or both?
5. Give TWO environmental factors that can affect your weight.

1. C 2. False 3. A and C 4. Both 5. Diet and the amount of exercise that you take.

Practice test questions

The following questions test levels 3-6

(a) The diagram shows part of a food web in a pond.

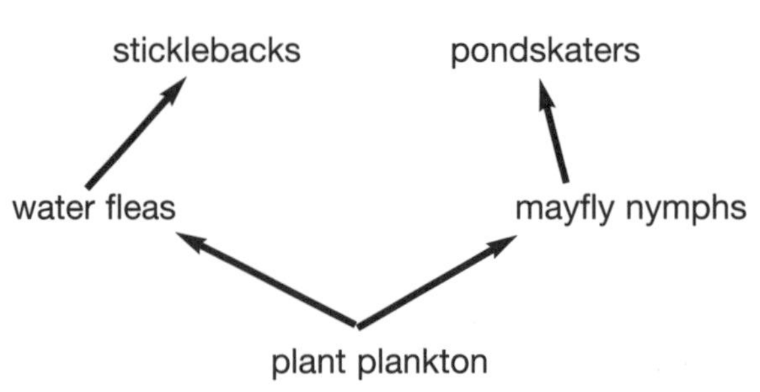

(i) What is the producer in this food web? **[1]**

....

(ii) How many of the organisms shown in the food web are animals? **[1]**

....

(iii) How many of the organisms shown in the food web are fish? **[1]**

....

(iv) A chemical is used to kill the plant plankton.

Explain what happens to the numbers of the other organisms shown in the food web. **[2]**

....

....

(b) The diagram shows some features of a stickleback.

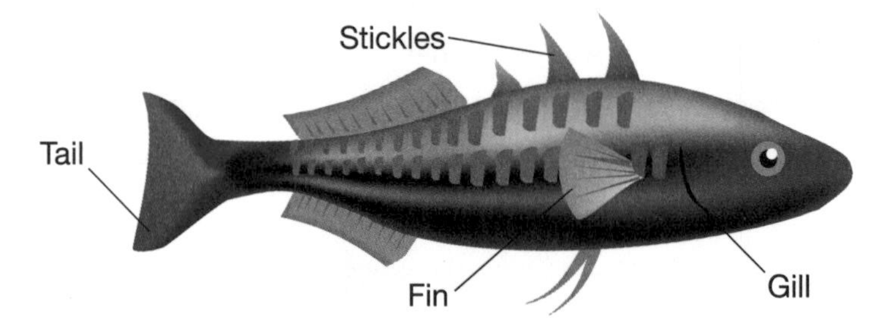

Explain how each of these features helps the stickleback to survive in its environment.

(i) Fin. **[1]**

....

(ii) Gill. **[1]**

....

(iii) Stickles. **[1]**

....

The following questions test levels 5-7

The diagram shows a cell.

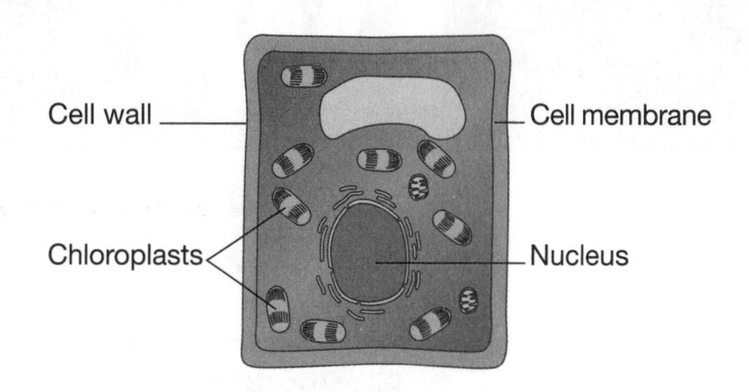

(a) How can you tell that the cell is from a plant and not from an animal?

Give two reasons. **[2]**

....

....

(b) Which of the labelled parts of the cell:

(i) controls all the activity of the cell? **[1]**

....

(ii) gives the cell a fixed shape? **[1]**

....

(iii) absorbs the energy needed for the plant to make food? **[1]**

....

(c) The diagram shows another cell from the plant.

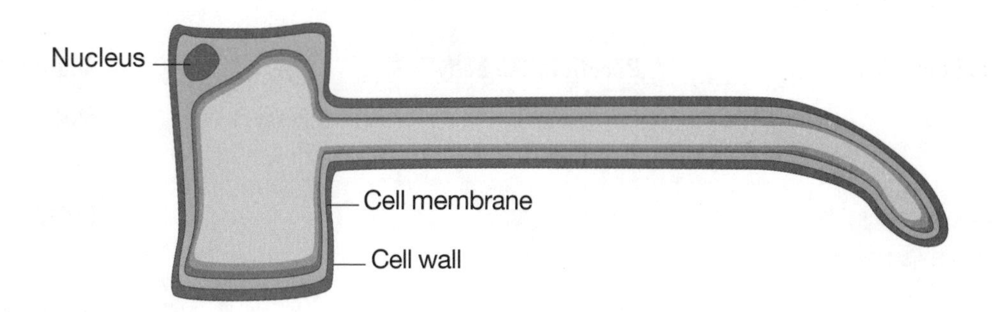

(i) Which part of the plant is this cell taken from? **[1]**

....

(ii) What feature of the cell enables it to do its job? **[1]**

....

Year 8

Chapter Two

The topics covered in this chapter are:

- **Food and digestion**
- **Respiration**
- **Microbes and disease**
- **Classification**

2.1 Food and digestion

After studying this topic you should be able to:

- **explain what is meant by a balanced diet**
- **describe chemical tests to identify proteins, carbohydrates and fats**
- **describe what happens to food in the digestive system**

A balanced diet

People often take the word diet to mean food necessary to lose weight. We all have a diet. Your diet is the food you eat.

A balanced diet provides:

- everything required for growth of the body
- everything required for repair of the body
- enough energy for the body's activities

The human diet must contain **proteins**, **carbohydrates**, **fats**, **vitamins**, **minerals**, **fibre** and **water**.

The table gives information about these different food chemicals.

Food chemical	Benefit to the body	Source
Proteins	Provide amino acids for building and repairing the body	Meat, fish, milk, cheese
Carbohydrates includes sugar and starch	Provide energy	Bread, potatoes
Fats	Store energy, give insulation	Butter, oil and margarine
Vitamins e.g. vitamin C	Required in small quantities for good health	Fruit, vegetables
Minerals e.g. iron	Required in small amounts for good health	Fruit, green vegetables

In addition to the food chemicals in the table, the diet should contain water and dietary fibre (roughage).

- **Fibre** is not digested, but helps in the production of faeces and prevents constipation. There is also evidence that fibre in the diet helps to retain water in the gut cavity and reduces the risk of bowel cancer.
- **Water** acts as a solvent, transports substances and provides a medium where reactions can take place.

Vitamins

Each **vitamin** has a particular job controlling a vital process in the body. Only small amounts are required to ensure good health.

The table gives some of the common vitamins and the jobs they do.

Vitamin	Good source	Use
A	green vegetables, butter, egg yolk, fish oils	healthy skin and membranes, prevents night blindness
B complex	yeast extract, liver, wholemeal bread	various, particularly respiration
C	citrus fruits, blackcurrants, vegetables	healthy skin, resistance to colds
D	butter, egg yolk, made in the skin	helps make bones

We now know the importance of eating fruit and vegetables that provide a source of Vitamin C. Scurvy affected sailors on long voyages. The disease caused bleeding gums, weakness of muscles and, ultimately, death. It was prevented by giving sailors fruit juice to drink. Dr James Lind found this out in 1747. British ships going on a long journey took limes to prevent scurvy. This led to British sailors being called 'Limeys'.

The first vitamin to be identified was vitamin B1. In 1896 a Dutch doctor, Dr Eilkman, was looking for a cure for a disease called beriberi. He noticed that some chickens in the hospital had a similar disease. He then noticed that the condition of the chickens changed when their food was changed from polished rice (rice with the outer husks removed) to whole grain rice (which contained the husks). He and his colleague, Dr Grijns, concluded that it must be a chemical in the husk that prevented beriberi.

In 1906 they boiled up some husks in water and used the solution to cure a pigeon that was suffering from beriberi. In 1934 scientists were able to identify the chemical in the husks and called it vitamin B1. In 1937 scientists were able to make vitamin B1 in the laboratory without having to extract it from rice husks.

Mineral salts

The table gives some examples of mineral salts needed in the human body. Like vitamins, they are needed only in small amounts.

Element	Good source	Use in the body
Ca (calcium)	cheese, milk	bones and teeth
F (fluorine)	toothpaste	hardening tooth enamel
Fe (iron)	liver, green vegetables	part of haemoglobin in red blood cells
I (iodine)	table salt additive, sea food	thyroid gland
K (potassium)	green vegetables	nerve and muscle function
Na (sodium)	table salt	nerve and muscle function

Progress Check

1 What are the seven types of food chemical needed in a healthy diet?

2 A man is advised by his doctor to reduce the fat and to increase the fibre in his diet. Which form of potatoes in the list would be most suitable for him? Explain your choice.
Chipped potatoes, potato crisps, jacket potatoes, mashed potatoes, boiled potatoes.

3 Which foods are good sources of calcium?

1. Proteins, carbohydrates, fats, vitamins, minerals, fibre, water 2. Jacket potatoes – skins rich in fibre. Chipped potatoes and potato crisps are cooked in fat and fat is absorbed. Mashed potato has fat added. 3. Milk and cheese.

Testing for proteins, carbohydrates and fats

Testing for proteins

The **biuret** test is used to test for proteins.
The food is added to water. Then sodium hydroxide solution and copper(II) sulphate solution are added.
A mauve colour shows the presence of protein.

Testing for carbohydrates

There are two tests here – for starch and for simple sugars, such as glucose.
If iodine solution is added to a food containing starch, the solution goes dark blue-black.
If Benedict's solution is added to a food containing a simple sugar and the mixture is heated, a yellow or red-brown colour is shown.

Testing for fats

Ethanol is added to the food being tested in a test tube. The test tube is shaken. The clear liquid is poured into distilled water. If the solution goes milky, a fat is present.
This happens because fat dissolves in ethanol but not in water.

Progress Check

The table gives the results of food tests on three foods, A, B and C.

Test	A	B	C
Protein test	Blue solution	Mauve colour	Blue solution
Starch test	Blue-black colour	No blue-black colour	Blue-black colour
Simple sugar test	Blue solution	Blue solution	Red-brown solid
Fat test	Clear	Clear	Milky white

What do these tests tell you about A, B and C?

A – starch; B – protein; C– starch, sugar and fat

The digestive system

Enzymes such as amylase and lipase help break down foods during digestion. They are biological catalysts.

If the digestive system from mouth to anus was stretched out, it would be several metres long.

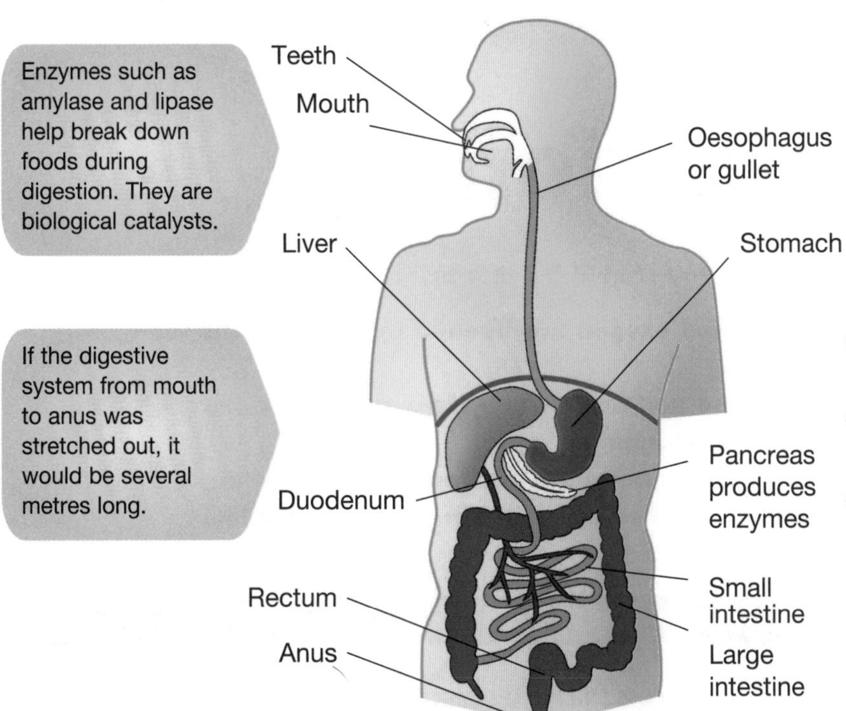

Fig. 2.1 The human digestive system.

Digestion is the breaking down of large insoluble food molecules into small molecules that the body can absorb and use.

The digestive process involves:

- a mechanical breaking down of food, e.g. by chewing
- a chemical breaking down using enzymes
- absorption of the small chemicals produced

The diagram shows the human digestive system.

The table shows what is happening at different places in the digestive system.

Place in the body	What is happening?
Mouth	Food chewed into small pieces for swallowing. It is mixed with saliva, which contains an enzyme. Enzyme (amylase) starts to break down starch.
Stomach	Food is stored for a few hours. Food is mixed with gastric juices. Hydrochloric acid kills most bacteria and gives the best conditions for enzymes (proteases) to break down proteins. In the small intestine alkaline juices neutralise stomach acids and make ideal conditions for breaking down fats. The food chemicals are absorbed here.
Large intestine	Water is removed and absorbed into the blood. Unusable material becomes faeces, which are stored in the rectum before being passed through the anus.

Absorption takes place in the small intestine. The food chemicals are absorbed into the bloodstream. They are transported to the liver. The liver controls what happens to the absorbed food chemicals.

- Glucose is needed for respiration. A constant supply of glucose circulates in the blood. The liver converts excess glucose into insoluble glycogen. This can be changed back into glucose when it is needed.
- Excess carbohydrates are converted into fats.
- Amino acids cannot be stored. They are used for growth and repair of the body. Excess amino acids are changed to urea, which is then excreted.

Progress Check

1 Which conditions, acidic or alkaline, are needed to breakdown (a) fats and (b) proteins?

2 Why is a regular supply of proteins needed by the body?

1. (a) alkaline (b) acidic 2. Proteins are broken down into amino acids. Amino acids are needed for the repair of the body. They cannot be stored so a regular supply is needed.

2.2 Respiration

After studying this topic you should be able to:

- **recall that glucose and oxygen are needed by human cells for respiration**
- **describe how glucose and oxygen are transported to cells**
- **explain the role of the lungs**
- **describe the differences between inhaled and exhaled air**
- **recall that respiration takes place in other living organisms**

Respiration in human cells

Respiration is the process that takes place in all cells to produce energy from glucose and oxygen. The glucose and oxygen reach the cells in the blood supply.

In the process of respiration, carbon dioxide and water are produced. Carbon dioxide is transported away from the cells in the blood.

Key Point

The word equation for respiration is:
glucose + oxygen → carbon dioxide + water

Respiration is a process that can be compared to burning.

The table compares the processes of respiration in cells with burning a carbon compound as fuel.

Respiration in cells	Burning a carbon compound such as fuel
uses food containing carbon, e.g. glucose	uses fuel containing carbon
uses oxygen	uses oxygen
produces carbon dioxide and water as wastes	produces carbon dioxide and water as wastes
releases energy, some as heat, some locked up	releases energy, most as heat, some as light

What is the role of the lungs?

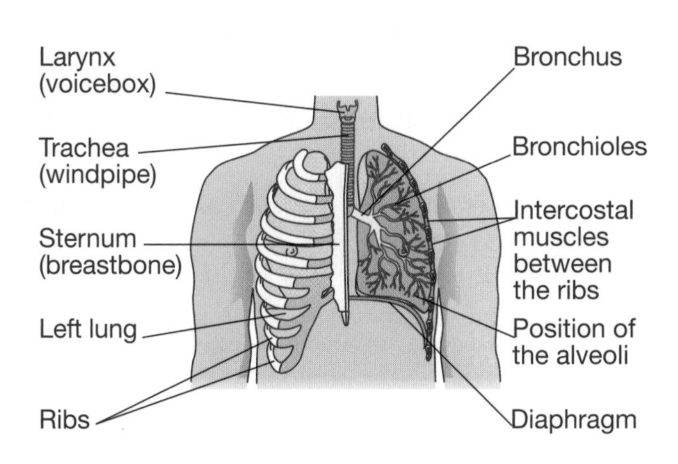

The diagram shows the human respiratory system. The lungs provide a means of getting oxygen into the bloodstream and removing the waste carbon dioxide.

Fig. 2.2 The human respiratory system.

Students frequently confuse respiration and breathing. Breathing is the mechanical process of inhaling and exhaling air. Respiration takes place in all the cells.

Air enters the lungs, which consist of branched tubes ending in millions of tiny sacs called **alveoli**. The walls of the alveoli are extremely thin and they have a very large surface area. Oxygen can diffuse through the alveoli into the blood and carbon dioxide can diffuse from the blood into the alveoli.

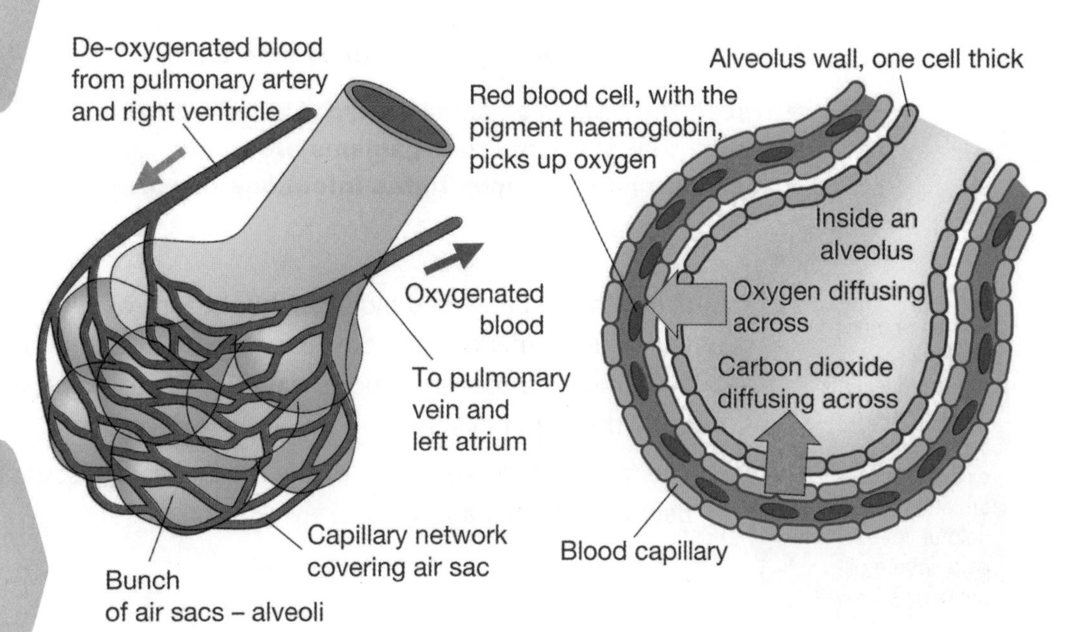

Students frequently forget that respiration takes place in plants because photosynthesis is so dominant in sunlight.

Fig. 2.3 The blood supply to air sacs.

Inhaled and exhaled air

The air we breathe out (called exhaled air) is different in composition from the air we breathe in (called inhaled air).

The table gives the typical approximate composition of inhaled and exhaled air.

Gas	Inhaled air	Exhaled air
Oxygen	21%	17%
Carbon dioxide	0.04%	4%
Nitrogen	79%	79%

Progress Check

1 How do the percentages of oxygen and carbon dioxide change between inhaled and exhaled air?

2 A person is having breathing difficulties. What treatment can be given to improve breathing?

3 A magazine article states that the percentage of oxygen in air in 1700 was 38%. Suggested why this might have reduced.

1. Oxygen decreases, carbon dioxide increases. NB Students often believe that all of the oxygen is removed. 2. Supply of oxygen. 3. Increased population, more combustion of fossil fuels, destruction of green plants.

2.3 Microbes and disease

After studying this topic you should be able to:

- **recall the names of different types of micro-organism**
- **recall that some micro-organisms are useful and some are harmful**
- **describe how the body fights infectious diseases**

There are other kinds of micro-organism. One is fungi. These too can be harmful, for example producing athlete's foot or helpful, for example as a source of antibiotics.

Ill health can be caused by a variety of micro-organisms, including bacteria and viruses entering the body. Bacteria can enter the body through ears, eyes, mouth, nose, anus, penis/vagina or a cut in the skin. Once in the body they can multiply.

The table summarises some of the similarities and some of the differences between bacteria and viruses.

Bacterium	Virus
Cell wall and cell membrane	Protein coat
Has genes but no nucleus	Has genes
Has cytoplasm	No cytoplasm
Can reproduce outside living cells	Can only reproduce inside living cells
Destroyed by antibiotics	Not destroyed by antibiotics
Bacteria produce toxins (poisons)	Viruses damage the cells in which they reproduce

How the body prevents micro-organisms entering

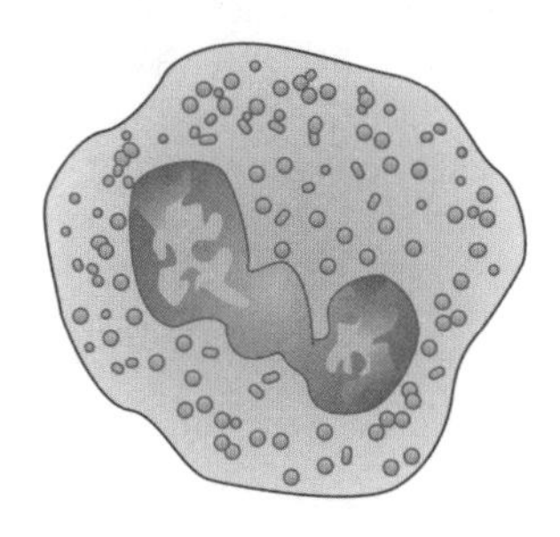

Fig. 2.4 White blood cell.

Students frequently write that white blood cells eat the microbes. This is not strictly correct – 'engulf' is the best word to use.

- The skin acts as a barrier.
- The breathing organs produce sticky mucus to trap microbes.
- Blood platelets produce clots to seal cuts.

White blood cells have a defensive role.

1 They engulf and destroy microbes.
2. They produce antibodies that destroy microbes.
3. They produce antitoxins that counteract the toxins produced by microbes.

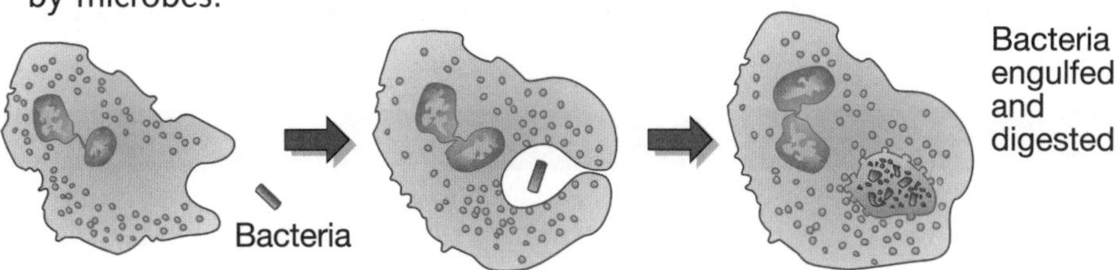

Fig. 2.5 White blood cell engulfing bacteria.

Once we have had a particular disease, the body knows how to produce antibodies and so the white blood cells can produce them quickly before the disease takes hold. We are said to be **immune** to that microbe.

Louis Pasteur (1822-1895)

In 1765, Lazzaro Spallanzani showed that food would not go bad if the microbes in it were destroyed. One way of killing the microbes in soup was to boil it.

Louis Pasteur was a French scientist who studied microbes and proved they were responsible for the process of decay.

Pasteur was able to isolate the bacteria that caused diseases such as cattle anthrax and chicken cholera.

Key Point

Bacteria are not all bad. They play an important role in many processes, for example sewage treatment.

Antibiotics were first discovered by accident. Sir Alexander Fleming noticed in 1928 that staphylococcus (a bacterium) in contact with a mould disappeared. He rightly concluded that there was an antibacterial substance in the mould. This was later found to be penicillin. Fleming did not succeed in making penicillin. That was 15 years later by HW Florrey and EB Chain. The ease with which penicillin can be destroyed is one reason why it was difficult to make. Since then the use of penicillin and other antibiotics has saved millions of lives.

Progress Check

1 Milk is pasteurised before being sold. What is this process and why is it done?
2 A doctor will not prescribe antibiotics to cure a virus. Why not?
3 Why is it important to take all of the antibiotics prescribed for a bacterial infection?

1. Milk is heated to a high temperature for a short period and then cooled. This kills bacteria.
2. Antibiotics have no effect on a virus. 3. Antibiotics kill bacteria. If the full course is not taken the few bacteria remaining can then multiply.

2.4 Classification

After studying this topic you should be able to:

- **recall how animals and plants can be classified**

Classification of animals

All living things can be divided into kingdoms. Two of these are the animal kingdom and the plant kingdom. Kingdoms are then divided in to phyla (singular phylum).

There are millions of different species of plants and animals and new species are being discovered all the time. Rather than try to study each one separately, it is sensible to put them together and study them as groups.

For example, if we consider the animal kingdom, we can divide all animals into two major groups:

- animals with backbones (called **vertebrates**)
- animals without backbones (called **invertebrates**)

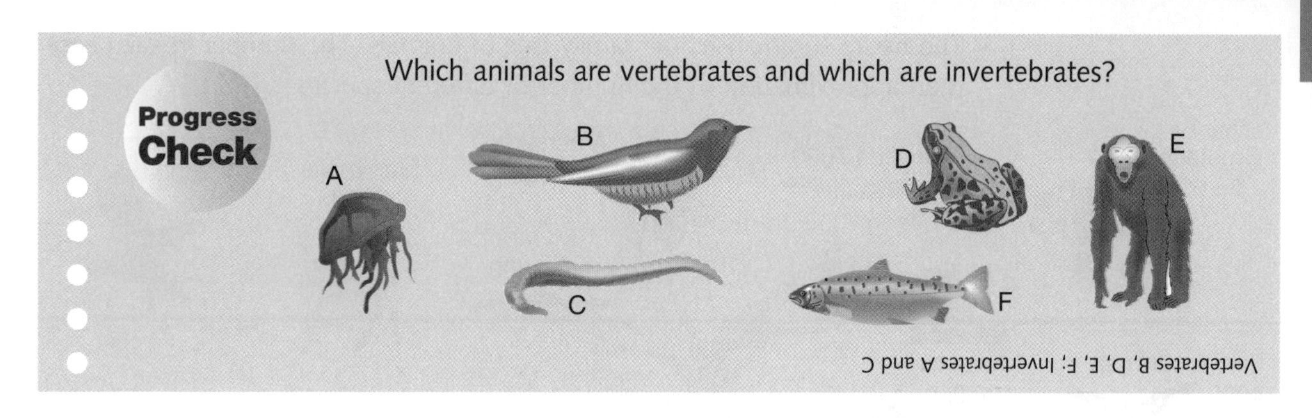

We can divide vertebrates further, according to whether the animal has a body temperature the same as its surroundings (called cold-blooded) or whether the animal is able to keep its body temperature the same on hot or cold days (called warm-blooded).

Cold-blooded animals cannot use internal processes to control their body temperature. However, a snake can warm up its body by basking in the sun.

The table lists common groups of vertebrates divided into cold-blooded and warm-blooded classes.

Cold-blooded	Warm-blooded
fish	birds
amphibians	mammals
reptiles	

Each of these classes has its own characteristics.

Fish Paired fins; gills.
Amphibians Slimy skin; spend some of their lives in water.
Reptiles Dry scaly skin; lay eggs on land.
Birds Feathers; lay eggs on land.
Mammals Hair; provide milk for young from special glands.

We can classify invertebrates in a similar way producing a number of different groups of phyla.

Phylum	Feature
Protozoa	made of one cell, e.g. amoeba
Sponges	animals made of similar cells loosely joined together
Cnidaria	body walls made of two layers of cells, e.g. jellyfish, sea anemones
Flat worms	flattened worm-like shape, e.g. tape worm
Annelida	worms made of segments, e.g. earthworms
Arthropoda	jointed legs, bodies made of segments; includes spiders, insects, centipedes
Mollusca	no segments; a fleshy pad on which they crawl, e.g. slug, snail
Echinodermata	star-shaped pattern, spiny skin, e.g. starfish

The figure summarises the family tree of animals. The number in each case is an approximation of the number of different species.

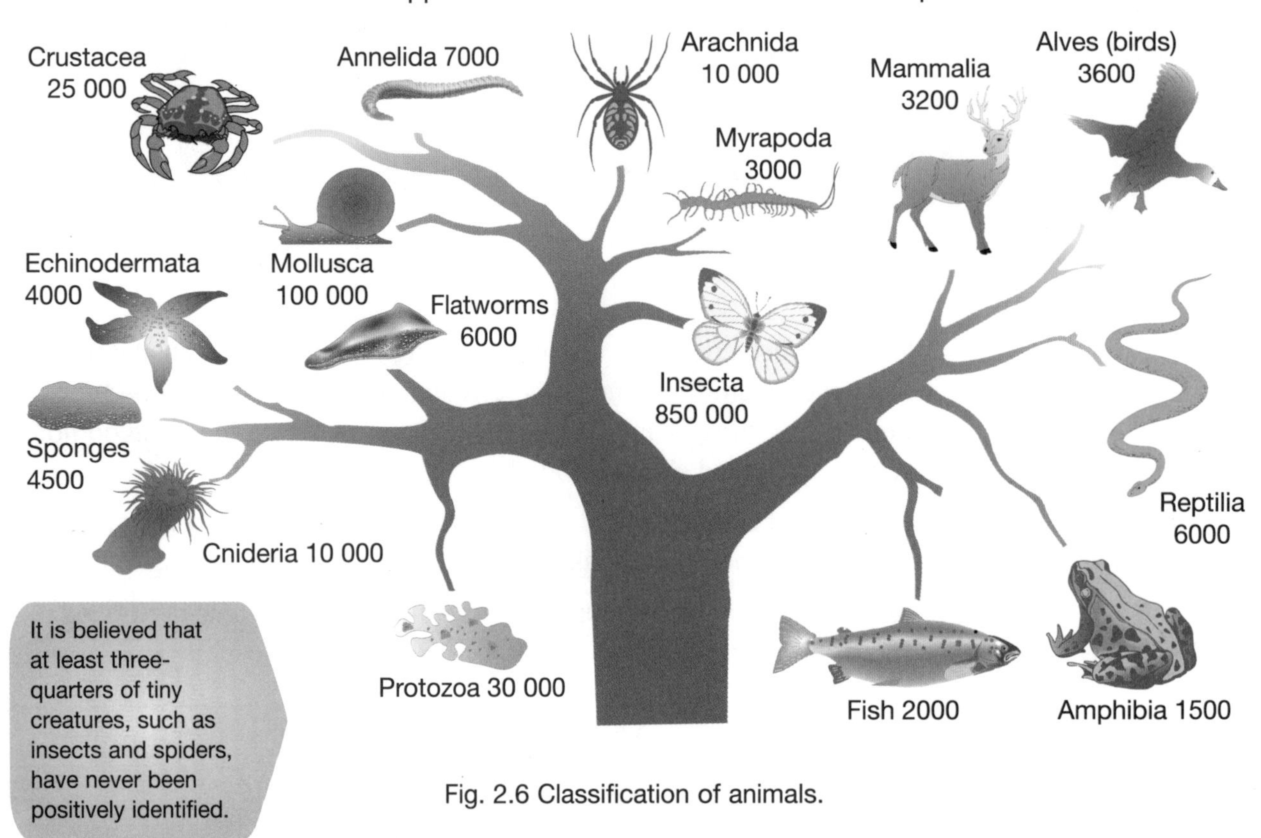

It is believed that at least three-quarters of tiny creatures, such as insects and spiders, have never been positively identified.

Fig. 2.6 Classification of animals.

Classification of plants

Plants can be classified in a similar way.
One way of dividing plants into classes is shown below.

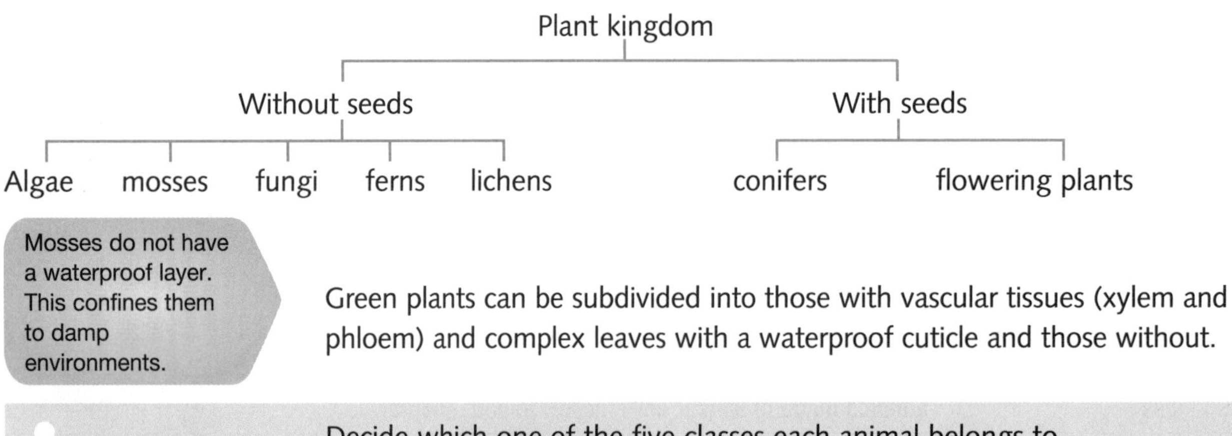

Mosses do not have a waterproof layer. This confines them to damp environments.

Green plants can be subdivided into those with vascular tissues (xylem and phloem) and complex leaves with a waterproof cuticle and those without.

Progress Check

Decide which one of the five classes each animal belongs to.

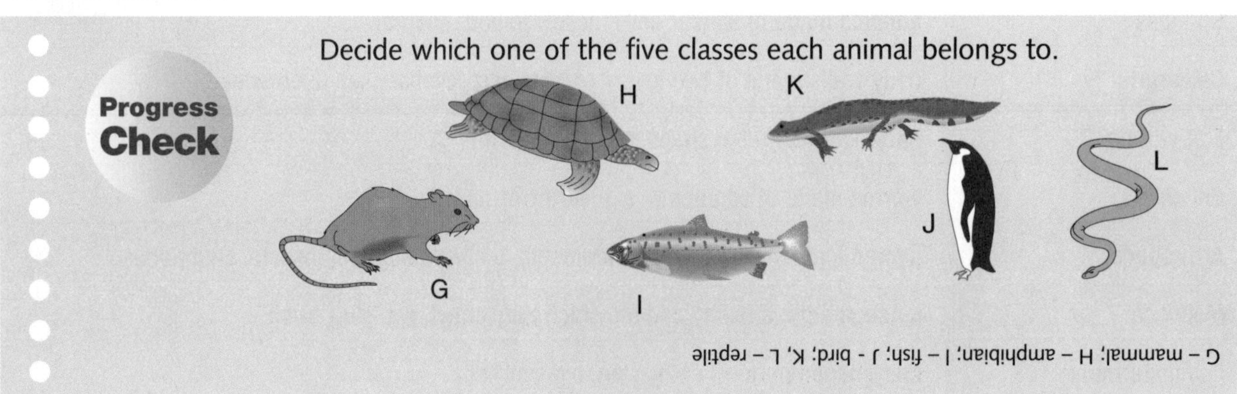

Practice test questions

The following questions test levels 3-6

The diagram shows some organs in the human body.

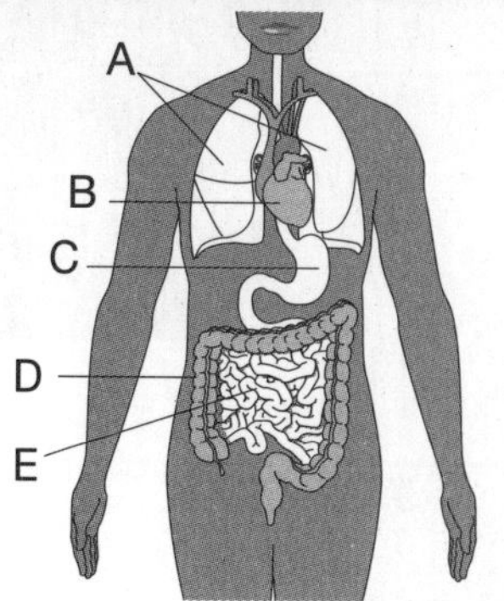

(a) Match up the name of the organ with the correct letter on the diagram. **[4]**

Stomach ____________. Lung ______________

Small intestine ____________ Large intestine ______________

(b) Write down the main function of each organ. **[4]**

Stomach ....

Lung ....

Small intestine ....

Large intestine ....

The following questions test levels 4-7

The table gives information about the composition of two brands of fruit yoghurt. The information was taken from the labels on the yoghurt pots.

	Brand X (per 100 g)	**Brand Y (per 100 g)**
Energy value	150 kJ	480 kJ
Protein	4.5 g	4.0 g
Fat	0.3 g	2.5 g
Carbohydrate	5.2 g	18.4 g
Additives	Preservative Artificial sweetener	Preservative

(a) Which brand of yoghurt would be more suitable as part of a slimmer's diet? Give a reason for your answer. **[3]**

....

....

(b) Brand Y contains no artificial sweetener. What could be used to sweeten this brand of yoghurt? **[1]**

(c) Why do the tests for protein, carbohydrate and fats not give any differences if used on these two yoghurts? **[1]**

....

(d) Calculate the mass of protein, fat and carbohydrate in one 125 g tub of brand Y. **[3]**

....

....

Chapter Three

Year 9

The topics covered in this chapter are:

- **Inheritance and selection**
- **Fit and healthy**
- **Plants and photosynthesis**
- **Plants for food**

3.1 Inheritance and selection

After studying this topic you should be able to:

- **recall that cells contain information that can be transferred from one generation to the next**
- **recall that genetic information from male and female is combined to produce a unique new individual**
- **describe the uses of selective breeding**
- **understand asexual reproduction and cloning**

Passing on information to the next generation

DNA is short for deoxyribonucleic acid.

All the cells in our bodies, except red blood cells, contain **nuclei**. The nucleus of a cell contains **chromosomes** that store information to enable new similar cells to be made and information to be passed onto future generations. The genetic information is stored on threads of **DNA**.

Most human cells contain 23 pairs of chromosomes. Twenty-two pairs are ordinary pairs. The other pair are sex chromosomes, which determine the sex of the person. If the sex chromosomes are alike (called XX) the sex is female, and if they are different (called XY) the sex is male.

After division by mitosis, each cell contains the same number of chromosomes as the parent cell.

When cells divide for growth or repair, they produce identical cells. This is by a process called **mitosis**. The diagram shows mitosis taking place in a cell with only four chromosomes.

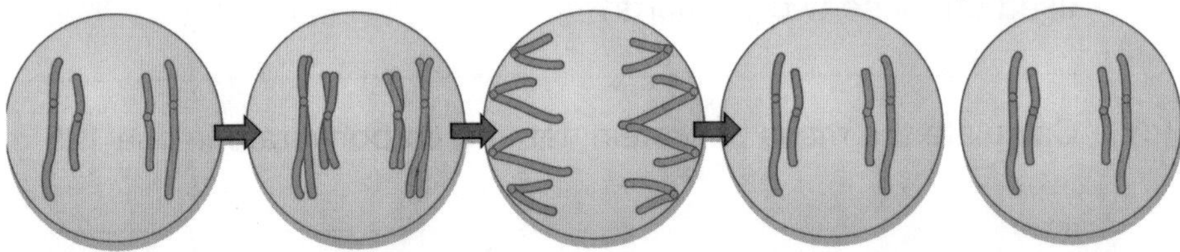

Fig. 3.1 Mitosis in a cell with only four chromosomes.

> Students often confuse mitosis and meiosis. Words such as 'meitosis' are used. The examiner cannot give credit for this word.

Sexual reproduction involves special cells called **sex cells**. These are egg cells in the female and sperm in the male. Sex cells are not produced by mitosis but by a process called **meiosis**.

During meiosis chromosomes still make an exact copy of themselves but the parent cell divides into four new cells. Each new cell has half the number of chromosomes of the original cell. The diagram shows how this happens, again with a cell containing four chromosomes.

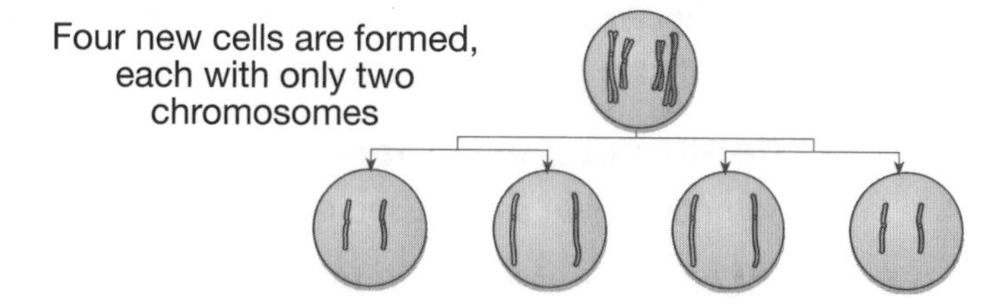

Fig. 3.2 Meiosis in a cell with only four chromosomes.

The male and female sex cells join together during fertilisation to produce a new cell (called the **zygote**), which has the characteristics of both parents and develops into the **embryo**. This is summarised in the diagram.

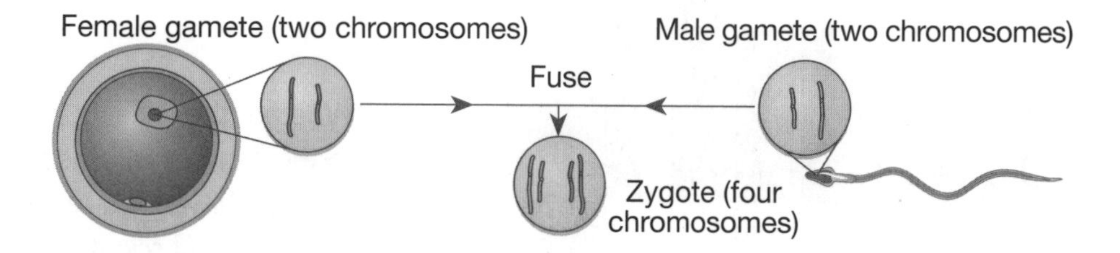

Fig. 3.3 Fertilisation produces a zygote.

The instructions carried by a chromosome for a particular characteristic, such as eye colour or blood group, are called **genes**. Every human being has two copies of each gene in every normal body cell; one is in each chromosome. One gene comes from the father and one from the mother.

Progress Check

A normal human cell contains 23 pairs of chromosomes.

1 How many chromosomes are there in a normal cell?

2 How many chromosomes are there in an egg cell?

3 How many chromosomes are there in a sperm?

4 How many chromosomes are there in a zygote?

1. 46; 2. 23; 3. 23; 4. 46

Variation

> This number is larger than the number of people who have ever lived.

> You will not have an exact double during your lifetime.

Have you noticed that brothers and sisters in the same family are not identical? Unless there are identical twins, the chances of parents having two identical children is about one in 1 800 000 000 000 000. This number is very large.

Variation is caused by different mixes of genes and by mutation.

There are two types of variation:

1 Discontinuous variation

This enables us to separate the population into different clearly distinguished groups, e.g. by blood group. We can sort the blood groups of individuals into four main groups:

A, B, AB and O. No one falls between two groups (e.g. a mixture of groups A and O).

Another example of discontinuous variation is albinism. This is a complete lack of skin pigment caused by the difference of a single gene.

2 Continuous variation

Sometimes we cannot see clearly different groups.

For example, if we were to measure the length of the middle finger of the right hand of 30 children, the results could not be clearly put into groups. Height and weight are other good examples.

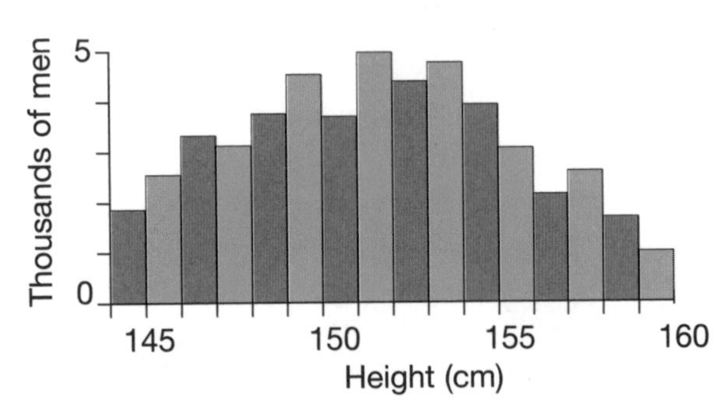

Fig. 3.4 Variation in the heights of a sample of men.

The graph shows the kind of variation that could be seen in the heights of a sample of men.

Whether the variation in height of these men is due to the genes they inherited from parents or the way the men have lived (environment) is a question that has interested scientists for many years.

In order to investigate this, scientists have studied identical twins. Identical twins have exactly the same genes because they are both formed from a single fertilised egg. This egg splits into two after it has been fertilised and two identical embryos are formed. So any differences in identical twins cannot be due to differences in their genes. Any differences must be due to the influence of the environment, such as amount and quality of food.

Studying identical twins that are separated shortly after birth and reared separately can give interesting information. If the twins still have similar characteristics, it suggests that inheritance was the main influence. If characteristics are very different, it suggests environment was the main influence.

Mutations

The copying of chromosomes when cells divide is very complicated and mistakes can occur. These mistakes are called mutations. Mutations can be caused by radiation and by some chemicals.

Most mutations are harmful but good mutations are possible.

Down's syndrome is caused by a mutation. Children with Down's syndrome have an extra chromosome. This occurs most frequently when the mother is older and the cell division to produce eggs has not occurred properly.

Progress Check

A study was carried out on a hundred pairs of identical twins. For 50 pairs of twins, each pair was brought up together. For the other 50, the pairs were separated at birth and brought up apart. Heights and weights were recorded at the same age. The results are shown in the table.

Difference in characteristic	Twins brought up together	Twins brought up apart
Average height in m	1.3	1.4
Average weight in kg	20	30

1 Look at the heights. Is there much difference between twins brought up together and separated? What does this suggest?

2 Is there much difference in weights? What does this suggest?

1. No, little difference. Suggests variation largely due to genetics.
2. Yes, big difference. Suggests variation due to differences in the environment.

Selective breeding

Farmers have always tried to produce better strains of plants and animals. The modern varieties of wheat bear little relationship to the wild grasses that were once cultivated. Farmers have tried to grow wheat that has shorter, sturdier stems with more seeds on each stem.

Selective breeding is slow and takes many generations.

A farmer wanting to produce chickens that grow quickly will choose parents that show this characteristic for breeding rather than slower growing chickens.

Asexual reproduction

Plants grown by collecting seeds will show variation because they involve genetic information from two parents. Plants grown by asexual reproduction are exactly the same and even inherit any fault in the parent.

Asexual reproduction involves only one parent and produces offspring that are identical to the parent.

Examples of asexual reproduction include:

- Single cell organisms such as amoeba and bacteria. They reproduce by growing and splitting into two identical halves.

Parts of a plant can grow into identical plants. Stem cuttings, leaf cuttings and so on will root and produce plants which are all identical to the original plant. The diagram shows the steps in producing new geranium plants from stem cuttings.

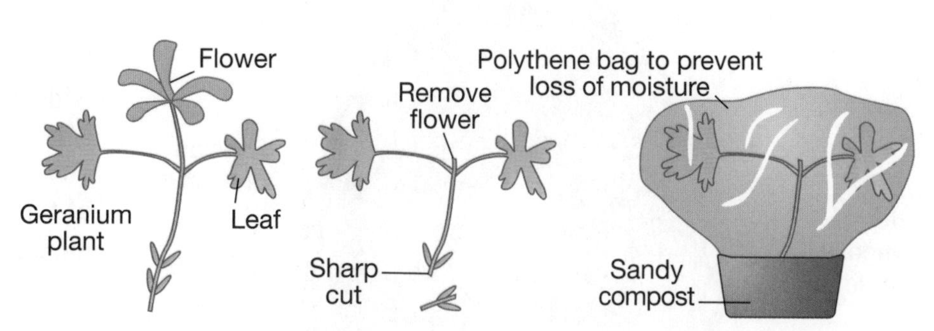

Fig. 3.5 Producing new geranium plants from stem cuttings.

- Dolly the sheep was produced by growing cells from the mother. She then possessed identical genetic material to the mother and was therefore identical to the mother. This process is called **cloning**.

Progress Check

1. Fruit trees have been grown by selective breeding that are only 2 metres high rather than 5–6 metres. What are the advantages to the farmer?
2. What are the advantages of growing wheat that has shorter, sturdier stems with more seeds on each stem?

1. Easier to pick. Less damage by wind. Easier to spray and prune. More trees in a given area.
2. Less easily blown over. Plants not wasting energy growing long stems. Higher yield.

3.2 Fit and healthy

After studying this topic you should be able to:

- **recall that the lungs, diaphragm, rib cage and associated muscles of the rib cage are essential for breathing**
- **describe the effects of smoking on the lungs and other body systems**
- **describe how alcohol and drugs affect the human body**
- **describe how simple joints function**

Fitness

There are two aspects of fitness:

1. The ability to perform certain physical tasks.
2. The speed of recovery after the activity.

Breathing

In Chapter 2.2 the process of respiration was considered. This involved oxidation of glucose in the muscles. The lungs were important in providing the oxygen to the bloodstream and removing the carbon dioxide.

The process of breathing is a mechanical process filling the lungs with air and then expelling waste air.

The diagram shows the structure of the chest cavity (called the **thorax**).

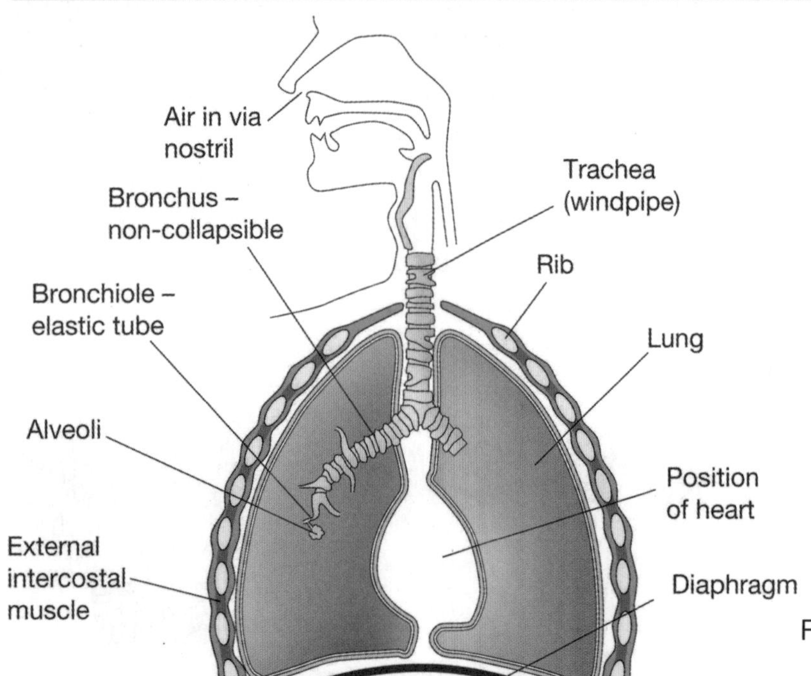

Fig. 3.6 Human respiratory pathway.

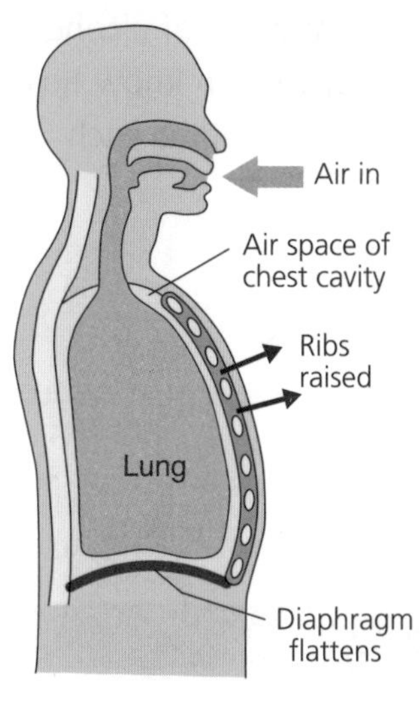

Fig. 3.7 Inhalation.

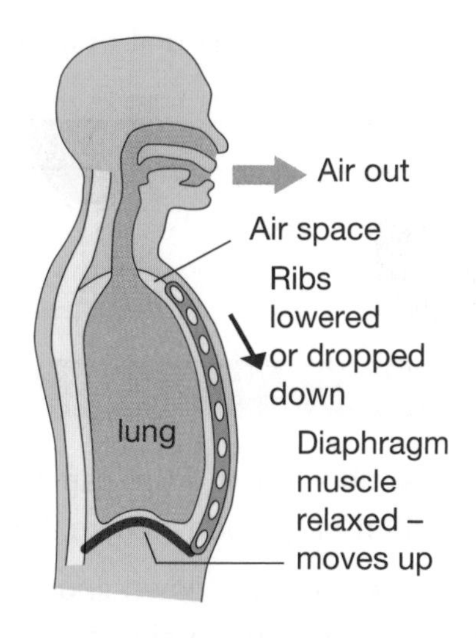

Fig. 3.8 Exhalation.

When you breathe in:

- the intercostal muscles move the ribs up and out
- the chest cavity becomes larger so the air pressure inside is reduced
- air rushes in to fill the lungs

When you breathe out:

- the intercostal muscles move the ribs down and in
- the chest cavity becomes smaller so the pressure increases
- air is forced out through the mouth to rescue the pressure

Effects of smoking

Tobacco contains a drug called nicotine. This drug speeds up the heart rate and raises blood pressure. Because of these changes a smoker has an increased risk of heart disease.

Other problems are caused when tobacco is smoked. The lungs have a mechanism to keep themselves clean. This involves producing a layer of mucus that is moved up and down the throat by moving hairs called **cilia**. Smoking slows down the movement of the cilia and produces more mucus. This collects in the bronchioles (tiny tubes in the lungs) causing a 'smoker's cough'.

Micro-organisms can get into the lungs more easily making diseases such as bronchitis more common. If this is not cleared up, permanent damage to the lungs can occur.

There are 50 million smokers in the United States of America. It is estimated that half of them will eventually die of smoking-related diseases.

The bronchioles become narrower and this makes breathing more difficult. The person has to breathe faster to receive the same amount of oxygen. This illness, called emphysema, will eventually be fatal. Smoking also increases the chances of lung cancer.

Even not smoking does not remove all of the risks of tobacco. It has been shown that breathing in smoke from a nearby smoker, known as 'passive smoking', can lead to certain health risks.

Effects of alcohol

When alcohol is swallowed it quickly gets into the bloodstream. In small amounts it can boost confidence and change personality. Alcohol is also a depressant drug that slows down the drinker's reactions. Co-ordination is clumsier and the vision can be impaired.

For these reasons there is a severe limit on the concentration of alcohol permitted in a car driver's blood. Unfortunately, no driver can know how much alcohol they can drink and still remain 'below the limit' as each human body uses up alcohol at different rates.

The following drinks contain one unit of alcohol.

Glass
of wine

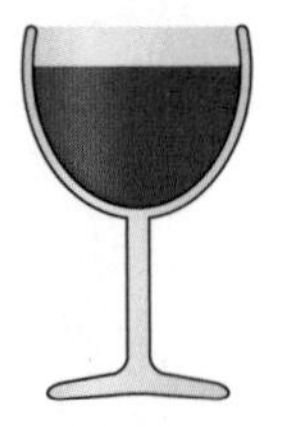

Single
whisky

Fig. 3.9 Drinks containing one unit of alcohol.

Effects of drugs

Like tobacco and alcohol, other drugs can affect the nervous system, which controls the operation of the body. There are four types of drug:

1. **Sedatives**, which slow the brain down and make the person sleepy.
2. **Stimulants**, which speed up the brain and make the person more alert.
3. **Hallucinogens** cause a person to have experiences that are different from real-life.
4. **Painkillers**, which remove our sense of pain.

Drugs can seriously affect health and can be addictive.

The working of simple joints

There are 206 bones in the human skeleton.

When two or more bones in the skeleton meet, a joint is formed which allows the movement of the skeleton. Bones in a joint are held together by strong fibres called ligaments. Muscles control movement at joints.

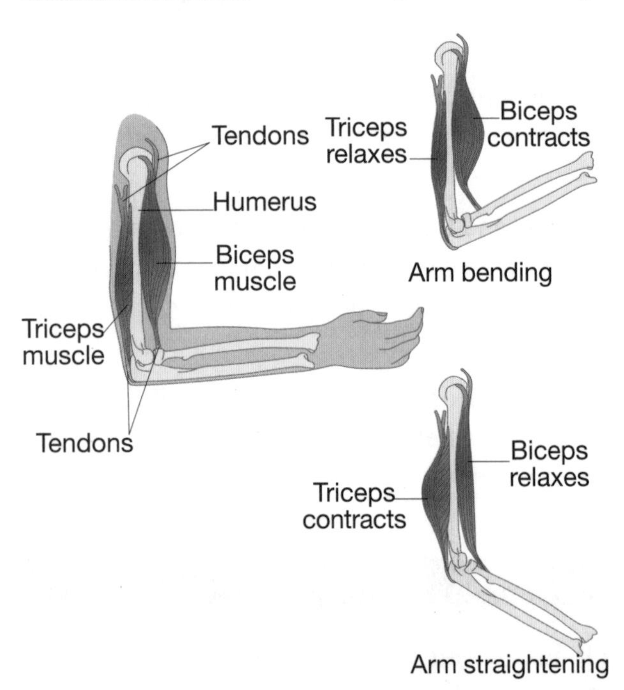

Fig. 3.10 Muscles in the upper arm.

Each end of a muscle is connected to a bone by a tendon, which will not stretch. One end of the muscle is attached to the bone that does not move, and the other is fixed to the bone that moves. The muscle pulls on the bone by contracting. This means the muscle gets shorter and fatter. When a muscle relaxes it gets longer but cannot exert a pulling force. Another muscle is needed to contract and pull the bone in the opposite direction.

Many bones in the body act as levers, which enable a relatively small force to lift a greater object. In the diagram the joint acts as a fulcrum. A small force applied on the contraction of the muscle in the forearm pulls on the bone. The far end of the bone that is carrying the weight is moved through a larger distance.

The diagram on page 40 shows two muscles, the triceps and the biceps. These work together to bend and straighten the arm. Muscles all over the body are arranged in pairs in a similar way. Because they work in opposite directions they are known as **antagonistic** muscle pairs.

Most joints in the body allow considerable freedom. These joints are called synovial joints. There are two types of **synovial joint**.

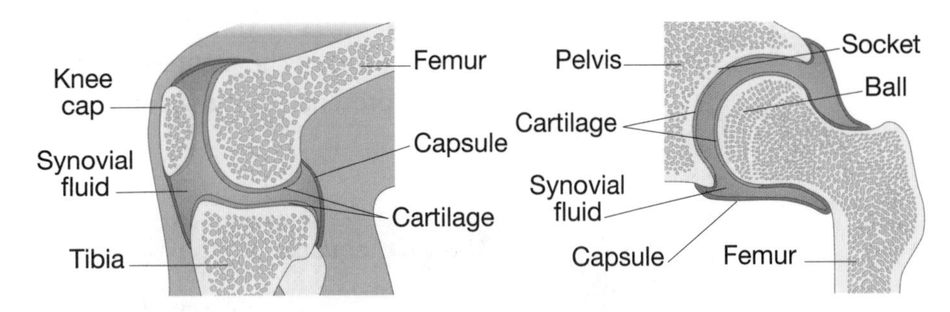

Fig. 3.11 Synovial joints.

Progress Check

1. The body uses up one unit of alcohol each hour. A person is drinking heavily until 2 am. Why is it unwise for the person to drive to work the next morning?
2. Which type of synovial joint allows movement in only one direction?
3. Which type of synovial joint allows movement in all directions?

1. The person may still be 'over the limit' in the morning. 2. Hinge joint 3. Ball and socket joint

3.3 Plants and photosynthesis

After studying this topic you should be able to:

- **recall that green plants produce glucose in the leaves by photosynthesis**
- **describe how leaf cells close to the upper surface are adapted for photosynthesis**
- **recall that some glucose produced during photosynthesis is stored as starch**
- **explain why green plants are important for the environment**

Photosynthesis

The diagram shows the parts of a plant.

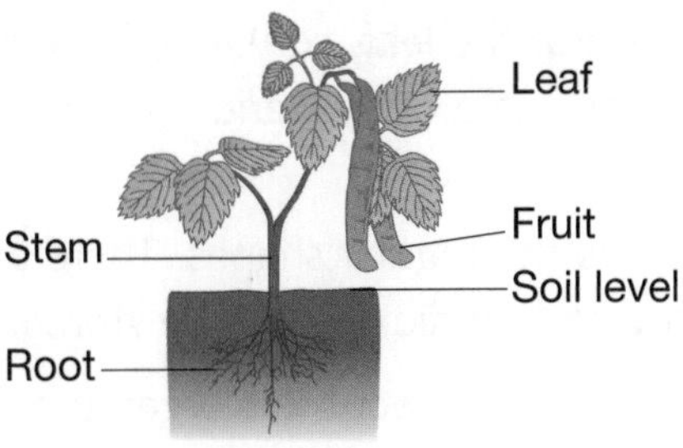

Leaf
Fruit
Stem
Soil level
Root

Fig. 3.12 Parts of a plant.

The roots anchor the plant in the soil. They also take water and important dissolved minerals from the soil. The water and dissolved minerals travel through hollow tubes in the stems to the leaves.

The leaves are the 'factories' of the plant. Here **photosynthesis** takes place, producing food for the plant.

Photosynthesis involves the reaction of water and carbon dioxide to produce glucose and oxygen. The oxygen is released into the atmosphere and the glucose is stored in the plant as starch. This process takes place in sunlight and in the presence of the green pigment, chlorophyll, which is a catalyst.

Key Point

The word equation for the reaction is

$$\text{Carbon dioxide} + \text{water} \xrightarrow{\text{chlorophyll}} \text{glucose} + \text{oxygen}$$

The word equation for photosynthesis is the reverse of the word equation for respiration.

Photosynthesis takes place in daylight. Respiration takes place all the time but is not noticed during the day.

Leaves usually have a large area to absorb the maximum amount of light. They are thin so that carbon dioxide does not have to travel far through the leaf. The veins in a leaf give some support and provide the leaf with a supply of water.

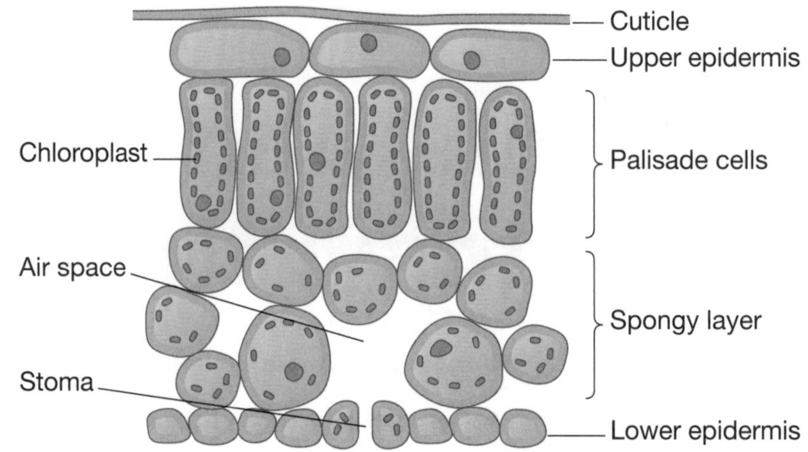

Fig. 3.13 Cross-section of a leaf.

The diagram shows the cross-section of a leaf. Carbon dioxide needed for photosynthesis enters through the **stomata** (singular **stoma**). Most of the stomata are on the underside of the leaf. Stomata open and close. Oxygen and water escape through the stomata. The stomata close at night to prevent too much loss of water.

The waxy layer on the surface of the leaf, called the **cuticle**, prevents evaporation of water from the surface. Below the cuticle there is a single layer of tightly-fitting cells called the **epidermis**. The **palisade cells**, below the epidermis but still near the surface of the leaf, contain a large number of **chloroplasts**. Chloroplasts contain the chlorophyll, and it is here that photosynthesis occurs.

The spongy layer contains large cells with irregular shapes. There are large gaps between the cells. Oxygen and carbon dioxide can be stored here.

Experiments showing photosynthesis often collect the oxygen produced.

Fig. 3.14 Experiment to show photosynthesis.

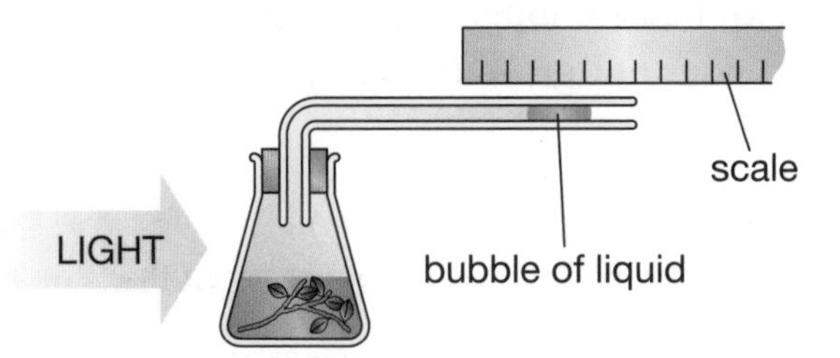

The apparatus in the diagram shows a gas being produced. Larger volumes of gas could be collected in a gas syringe.

Photosynthesis is an important process as it replaces oxygen in the atmosphere.

Progress Check

Finish the table by adding words from the list.

Palisade cells Spongy mesophyll cells Stomata Waxy cuticle

Part of leaf	What it does
	helps waterproof the leaf
	the holes through which gases get in and out
	most photosynthesis occurs here
	loosely packed cells that store gases

waxy cuticle, stomata, palisade cell, spongy mesophyll cells

3.4 Plants for food

After studying this topic you should be able to:

- **identify a range of minerals needed for healthy plant growth**
- **recall that a population of predators influences the number of prey organisms**
- **explain the advantages and disadvantages of using pesticides**
- **explain the advantages and disadvantages of growing crops in a greenhouse**

Minerals needed for plant growth

Plants need large amounts of nitrogen, phosphorus and potassium if they are to grow well. As plants grow they use up supplies of these minerals. They can be added using fertilisers.

The table shows how these three elements are used in the growing plant.

Element	Importance of the element to a growing plant
Nitrogen, N	Necessary for the growth of stems and leaves
Phosphorus, P	Essential for root growth
Potassium, K	Needed for the production of flowers

Other elements, including magnesium, sulphur and calcium, are needed in smaller amounts. Some elements, called **trace elements**, are needed in very small amounts.

Magnesium is needed for photosynthesis. If soil does not have enough magnesium in it, the plant cannot make enough chlorophyll. The plant is able to absorb less sunlight and does not grow. It does not make enough food and gives a low yield.

Progress Check

1. A fertiliser bag is labelled NPK 15:5:10. What does that suggest about the fertiliser?
2. A fertiliser contains sodium nitrate ($NaNO_3$) and potassium phosphate (K_3PO_4). Why is this a good general fertiliser?

1. 15 parts nitrogen, 5 parts phosphorus and 10 parts potassium. 2. It contains nitrogen, phosphorus and potassium to help with growth of stems, roots and flowers.

Predator–prey relationships

A **predator** is a carnivorous animal. The animal or animals it eats are called **prey**. A fox (predator) eats rabbits (prey). The graph shows how the numbers of foxes and rabbits change over a period of time.

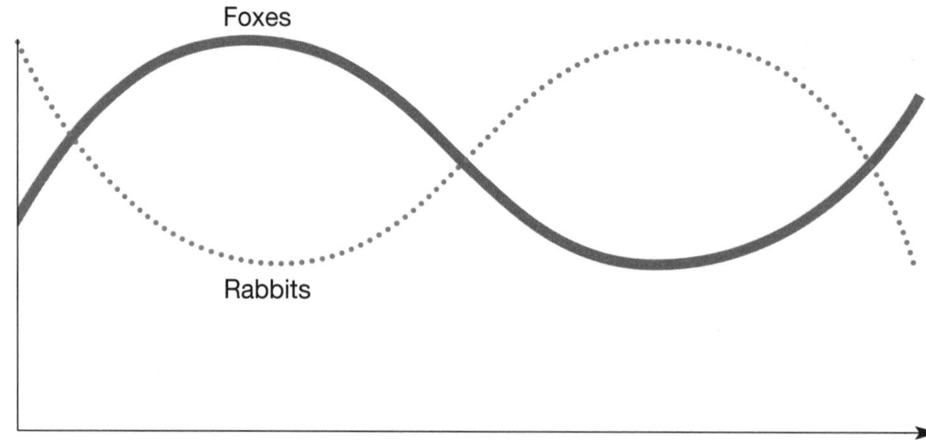

Fig. 3.15 How the numbers of foxes and rabbits change over time.

Notice that when there are few foxes the number of rabbits increases and when there are a large number of foxes the number of rabbits decreases.

Pesticides

Pesticide sprayed onto plants
↓
Pesticide kills insects
↓
Insects eaten by a small bird
↓
Bird of prey eats smaller birds
↓
Bird of prey lays eggs

Chemicals can be used to kill pests. These chemicals are called pesticides. However, chemicals can have serious side effects. DDT is a pesticide made from chlorine. It was used widely for killing pests about 50 years ago.

DDT stands for dichlorodiphenyl trichloroethane

The DDT does not break down. It builds up in the insects and then in the fat of the birds who eat them. It was noticed that there was a big drop in the number of birds of prey. It was found that DDT made the shells of the eggs of birds of prey very thin. These would break easily without hatching. Now DDT and similar pesticides are banned in Great Britain.

The diagram shows how the DDT concentrations build up in different trophic levels. The concentrations are in parts per million (ppm).

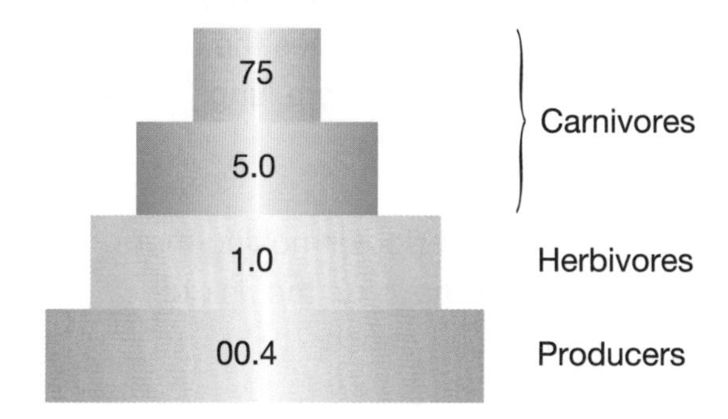

Fig. 3.16 DDT concentrations.

Advantages and disadvantages of growing in a greenhouse

The table gives some advantages and disadvantages of growing plants in a greenhouse.

Advantages	Disadvantages
Higher temperature than outside	Pests and diseases can build up
Crops grown earlier and later than outside	Costs of greenhouse
Controlled conditions	Running costs
Less wind damage	
Increasing carbon dioxide increases rate of photosynthesis	

Farmers often use polytunnels to grow plants to yield early crops. Polythene sheeting stretched across a frame gives protection to growing plants.

Progress Check

1. In 1965 DDT was detected in the livers of penguins in Antarctica. Suggest how DDT used in farms in Great Britain could eventually find its way into these penguins.
2. Why does DDT cause more problems in birds of prey than in small animals, such as caterpillars and small birds?

1. Pesticides get washed into rivers and the sea; absorbed by algae; eaten by small fish; eaten by larger fish; eaten by penguins. 2. DDT builds up in bigger animals as they eat more prey.

Practice test questions

The following questions test levels 3-6

(a) The graph shows how the number of cigarettes smoked changes the risk of the death of a person from lung cancer.

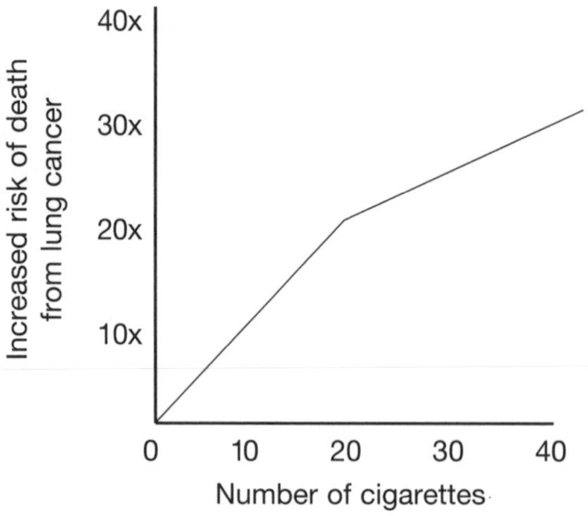

Increased risk of death from lung cancer
40x
30x
20x
10x
0 10 20 30 40
Number of cigarettes

(i) What effect does smoking more cigarettes have on the risk of getting lung cancer? **[1]**

....

(ii) What is the increased risk of dying from lung cancer if a person smokes 10 cigarettes a day? **[1]**

....

(iii) Why would you suggest that a friend who cannot give up smoking should reduce the number of cigarettes smoked? **[1]**

....

....

(b) Explain how cigarette smoke reduces the exchange of gases in the lungs. **[3]**

The following questions test levels 4-7

(a) Mary wants to produce some new geraniums. She has some tall, spindly plants from the previous year. She looks in a catalogue and finds she can buy geranium seeds.

Write down two advantages and two disadvantages of making new plants from cuttings over growing new plants from seed. **[4]**

....

....

....

(b) Some of the geraniums have variegated leaves. These have patterns of yellow and green on them.

(i) Why does photosynthesis only take place on the green parts of the leaf and not the yellow? **[2]**

....

....

(ii) Write down the word equation for photosynthesis. **[2]**

....

Chemistry

Chapter Four (Year 7)		Studied	Revised	Practice Questions
4.1 Acids and alkalis	What are acids and alkalis Detecting acids and alkalis Neutralisation			
4.2 Simple chemical reactions	Making new materials Metals and acids Carbonates and acids Burning			
4.3 Solids, liquids and gases	Classify materials as solids, liquids and gases Using a particle model to explain observations			
4.4 Solutions	Making solutions Separating mixtures Particle model to explain dissolving			
Chapter Five (Year 8)				
5.1 Atoms and elements	Elements			
5.2 Compounds and mixtures	Elements combining to form compounds Common mixtures			
5.3 Rocks and weathering	How old are the rocks of the Earth? What are rocks made from? Weathering			
5.4 The rock cycle	Types of rock Rock cycle			
Chapter Six (Year 9)				
6.1 Reactions of metals and metal compounds	Uses of metals Reactions of acids Salts			
6.2 Patterns of reactivity	Reactions with water Reactions with air or oxygen Reactivity series Predicting reactions			
6.3 Environmental chemistry	Acid rain Monitoring air and water pollution Global warming			
6.4 Using chemistry	Energy from fuels What makes a good fuel? Hydrogen as a fuel Energy from other chemical reactions Mass changes during a reaction			

The topics covered in this chapter are:

- **Acids and alkalis**
- **Simple chemical reactions**
- **Solids, liquids and gases**
- **Solutions**

4.1 Acids and alkalis

After studying this topic you should be able to:

- **name some common acids and alkalis**
- **recall that solutions can be classified as acidic, alkaline or neutral using indicators**
- **understand the pH scale and recall what happens to pH when a substance is neutralised**
- **know some everyday uses of acids, alkalis and neutralisation**

What are acids and alkalis?

The acids you use will probably look like water as they are mixed with water or diluted before use. Never assume that a colourless liquid is water.

Many substances in the world around us are acids or alkalis. The sharp taste we get when we bite into an apple is an acid. Acids always have a sour taste although we would be very unwise to taste most of them!

Acids are found in lemons, oranges and limes. These are called citrus fruits and the acid is citric acid. The sourness of vinegar is caused by the acid it contains. This is called ethanoic acid.

There are three common acids we use in the laboratory. They are sometimes called mineral acids. They are:

Sulphuric acid	H_2SO_4
Nitric acid	HNO_3
Hydrochloric acid	HCl

Alkalis are in some ways the opposite of acids. They will react with and 'cancel out' acids.

Common alkalis in the home are found in bicarbonate of soda, washing powders and washing soda. Common alkalis in the laboratory include:

Sodium hydroxide	$NaOH$ (sometimes called caustic soda)
Potassium hydroxide	KOH
Calcium hydroxide	$Ca(OH)_2$
Ammonia	NH_3

Substances, such as water, that are neither acidic nor alkaline are said to be **neutral**.

Progress Check

1. Write down the names of five acids.
2. Write down the names of five alkalis.
3. All acids are made up of different elements. Sulphuric acid is made up from hydrogen, sulphur and oxygen. Which elements make up nitric acid and hydrochloric acid?
4. Which element is present in all acids?

1. Citric acid, ethanoic acid, sulphuric acid, nitric acid, hydrochloric acid.
2. Bicarbonate of soda, washing soda, washing powders, sodium hydroxide, potassium hydroxide, calcium hydroxide, ammonia. 3. Nitric acid – hydrogen, nitrogen and oxygen. Hydrochloric acid – hydrogen and chlorine. 4. Hydrogen

Fig. 4.1 Hazard warning sign on a bottle.

Look for hazard warning signs on containers at home, such as bleach or washing powder.

Substances containing acids or alkalis often have a warning or **hazard sign** on them. Here are three warning signs you might see on bottles:

Harmful
Can make you ill if swallowed, breathed in or absorbed through the skin.

Irritant
May cause reddening or blistering of the skin.

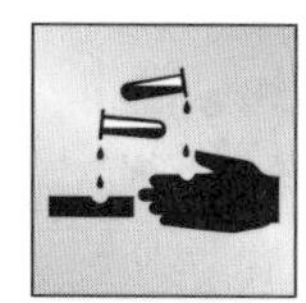

Corrosive
Attacks and destroys living tissue.

Fig. 4.2 Hazard warning signs.

Acids and alkalis have to be transported from the factory where they are made to places where they are going to be used. They are often transported in road tankers. There are hazard warning signs on these tankers. These warnings help the emergency services if they have to attend an accident.

When you are out, look for hazard signs on road tankers.

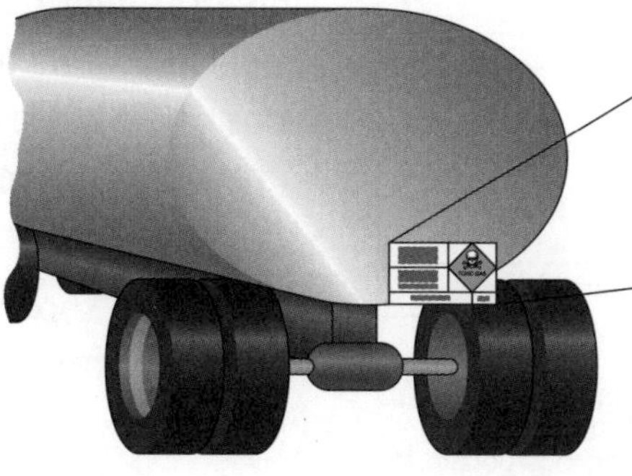

2PE

1830
Sulphuric acid

TOXIC GAS

Teeside (0099) 12345

Haz-Chem Ltd.

This label tells the emergency services that the tanker contains sulphuric acid. The 2PE is a code and it tells the emergency services how to deal with the chemical in the tanker.

Fig. 4.3 Hazard sign on a tanker.

Detecting acids and alkalis

Acids and alkalis can be detected using coloured solutions obtained from plants. These coloured substances change colour when added to acids and alkalis. They are called **indicators**.

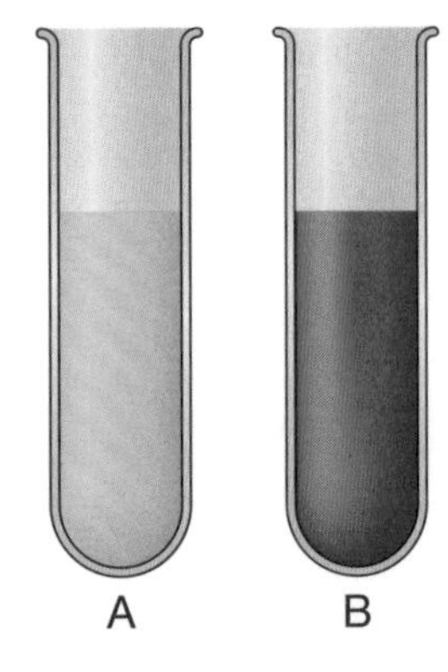

Fig. 4.4 Which test tube contains an alkali?

If some red colour is extracted from red cabbage it can be used as an indicator.

If the red cabbage solution is added to an acid, the solution stays red. If it is added to an alkali it turns green.

Other extracts from plants can be used. Litmus is a purple-coloured extract from a lichen or moss. If **litmus** is added to an acid, the solution turns red. If it is added to an alkali, the solution turns blue.

Sometimes scientists use a piece of paper soaked in litmus. They call this litmus paper. It is easier than carrying bottles of liquid.

Key Point

Acid	Alkali
Red	Blue

Progress Check

1 What colour is litmus in a neutral solution?

(Hint: Think of neutral as being halfway between acid and alkali.)

2 Three test tubes contain different liquids. A piece of red litmus paper and a piece of blue litmus paper are added to each tube. The results are:

Liquid	red litmus	blue litmus
A	stays red	turns red
B	turns blue	stays blue
C	stays red	stays blue

What can you conclude about each liquid from these tests?

1. Purple 2. A – acid; B –alkali; C – neutral.

	pH	Colour of Universal Indicator	Examples in the home	Examples in the laboratory
STRONG ACIDS	1		car battery acid	mineral acids
	2			
	3	red		
	4		lemon juice, vinegar	ethanoic acid
WEAK ACIDS	5	orange		
	6	yellow	soda water	carbonic acid
NEUTRAL	7	green	water, salt, ethanol	
WEAK ALKALIS	8	blue	soap, baking powder	sodium hydrogencarbonate
	9	blue-purple		
STRONG ALKALIS	10			ammonia solution
	11		washing soda	
	12	purple	oven cleaner	
	13			sodium and potassium hydroxides
	14			

Fig. 4.5 The colours for a simple form of Universal Indicator.

Litmus paper and similar indicators are useful for detecting acids and alkalis but they do not compare their different strengths. **Universal Indicator** is better. It is a mixture of simple indicators and it changes through a series of different colours. The pH value can be found from the colour.

The pH value is a number on a scale from 0 to 14 which shows how acidic or how alkaline a substance is. The indicator shows a different colour for different **pH values**.

If Universal Indicator solution is added to a solution and the solution turns orange, the pH value is 5 and the substance is a weak acid.

Progress Check

1. What is the pH value of a neutral solution?
2. What colour does Universal Indicator turn in a solution of pH value 9?
3. What pH value has a solution that turns Universal Indicator yellow?
4. Two solutions are tested with Universal Indicator. Solution X goes red and solution Y turns yellow. Which is the stronger acid?

1. 7 2. Blue 3. 5 4. X

Scientists wanting to make a number of pH readings may use a **pH meter**. This consists of a glass probe attached to a meter. A reading can be made or pH values transferred to a computer.

Fig. 4.6 A pH meter being used to measure the pH of a solution.

Neutralisation

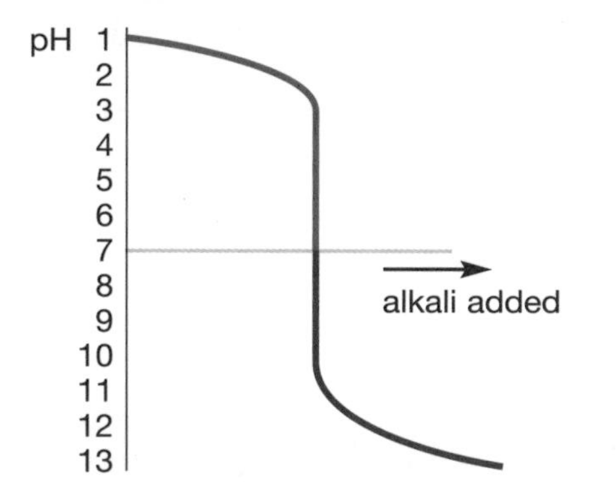

Fig. 4.7 Graph showing changes in pH.

When an alkali is added to an acid, the pH value changes. The graph shows the changes when sodium hydroxide is added to hydrochloric acid. A pH meter is used to follow changes in pH.

The effects of the acid are cancelled out by the alkali. This change is called **neutralisation**.

Examples of neutralisation

1. Everyone has several hundred cubic centimetres of hydrochloric acid in the gastric juices of the stomach. This is used in the digestion of food. Minor problems of indigestion can be caused by excess acid in the stomach. This excess acid can be neutralised by adding a weak alkali. This is called an **antacid**. Suitable antacids are milk of magnesia (a suspension of magnesium hydroxide) and bicarbonate of soda (sodium hydrogencarbonate).

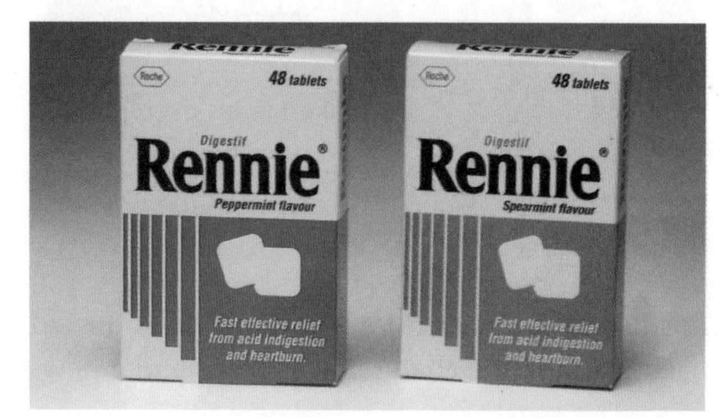

Fig. 4.8 Antacids.

2 Farmers have to control the pH values of their soil. If the soil becomes too acidic, a good yield of crops cannot be grown. Rain and artificial fertilisers make the soil more acidic. The farmer can neutralise the excess acidity by treating the soil with lime (calcium hydroxide).

Progress Check

1 If some sulphuric acid is spilt onto the floor, bicarbonate of soda is added. Why is this done?

2 A wasp sting is treated with vinegar but a bee sting is treated with bicarbonate of soda. Suggest why different treatments are given.

3 Inland lakes can become too acidic for fish to live in. What can be added to the lake to neutralise the water?

1. To neutralise the acid. 2. Wasp sting contains alkali so needs an acid to neutralise it. Bee sting contains an acid so needs an alkali. 3. Lime

4.2 Simple chemical reactions

After studying this topic you should be able to:

- **understand that new materials are formed during a chemical reaction**
- **recall what happens when metals react with acids, when acids react with carbonates and when materials burn**
- **describe tests for hydrogen and carbon dioxide**
- **describe burning as a reaction with oxygen**

Making new materials

Bricks have been made for thousands of years using very similar methods. The steps used to make bricks are:

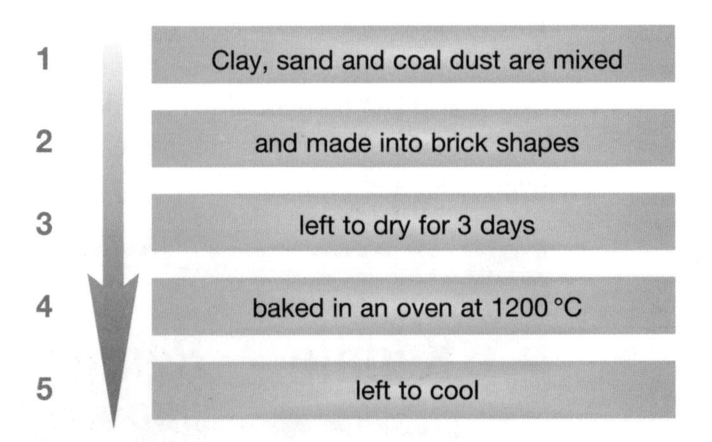

In stages 2–4 the mixture is soft and could be cut with a knife. During the baking a **chemical** change takes place and a new hard material is produced.

Key Point

New materials are produced after a chemical reaction. It is impossible to turn a brick back after it is baked.

Metals and acids

Key Point

Many metals will react with dilute acids to produce hydrogen gas.

If a piece of magnesium is added to dilute hydrochloric acid, bubbles of colourless hydrogen gas are seen. This reaction can be summarised as:

magnesium + hydrochloric acid → magnesium chloride + hydrogen

If a lighted splint is held close to the mouth of the test tube, the hydrogen gas burns with a squeaky pop.

Carbonates and acids

Key Point

A carbonate reacts with a dilute acid to form carbon dioxide gas.

If some calcium carbonate is added to dilute hydrochloric acid, bubbles of colourless carbon dioxide gas are seen. This reaction can be summarised as:

calcium carbonate + hydrochloric acid → calcium chloride + water + carbon dioxide

The cloudy white colour of limewater looks milky.

The gas is bubbled through colourless limewater solution. The solution goes cloudy white.

Progress Check

Here is a list of chemicals:

carbon dioxide   copper oxide   hydrogen   sodium carbonate   zinc

1 Which chemical reacts with dilute sulphuric acid to form hydrogen gas?

2 Which chemical reacts with dilute sulphuric acid to form carbon dioxide gas?

3 Which gas can be tested for with a lighted splint?

4 Which gas is tested for with limewater?

1. Zinc 2. Sodium carbonate 3. Hydrogen 4. Carbon dioxide

Burning

Key Point

When a material burns it forms oxides.

Combustion is another name for burning.

A material uses up oxygen when it burns.

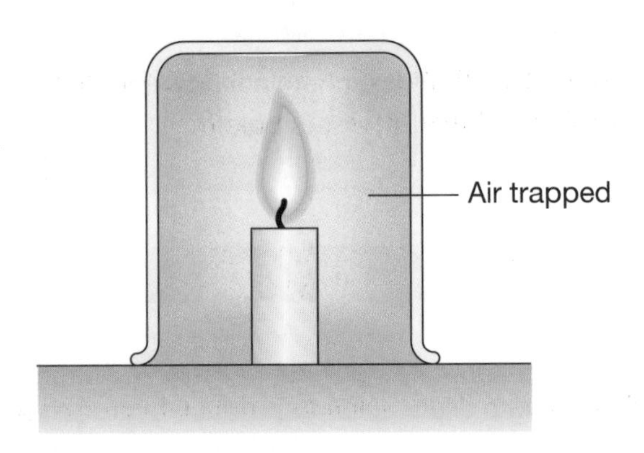

Fig. 4.9 A candle burning under a beaker.

The candle continues to burn under the beaker as it uses up the oxygen. The candle goes out when most of the oxygen is used up. The candle forms a number of products when it burns, including carbon dioxide and hydrogen oxide (water).

Flame will go out if oxygen is cut off.

Metals often burn in air or oxygen to form oxides. For example, a piece of magnesium burns in oxygen to form magnesium oxide.

magnesium + oxygen → magnesium oxide

Progress Check

1 What is formed when zinc burns in oxygen?
2 Which gas is used up when burning takes place?
3 Why does the mass of magnesium increase when it burns in oxygen?

1. Zinc oxide 2. Oxygen
3. Oxygen from the air joins with magnesium to form magnesium oxide.

4.3 Solids, liquids and gases

After studying this topic you should be able to:

- **classify materials as solids, liquids or gases**
- **explain that all substances are made of particles**
- **describe the movement and arrangement of particles in solids, liquids and gases**
- **use the particle model to explain some observations**

Classify materials as solids, liquids and gases

Materials can be classified as solids, liquids or gases. However, all materials are made up from tiny particles.

The word material is often used to describe a type of fabric, such as cotton. In Science it is used to describe anything around you. Clay, water and air are all materials.

Particle model of solids, liquids and gases

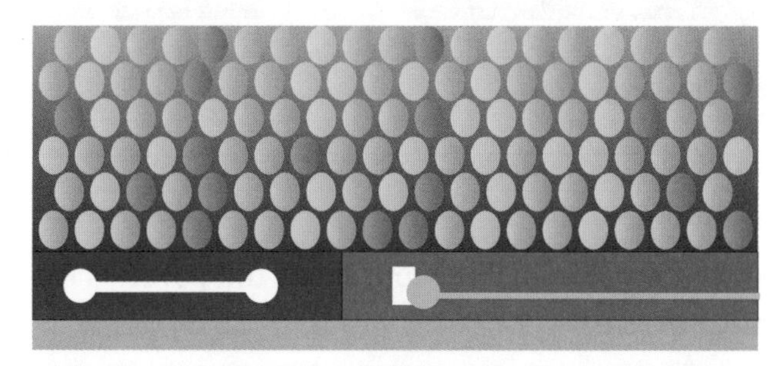

Fig. 4.10

Look at the picture of the crowds in the stand at a football match. The regular arrangement of the seats means there is a pattern in the way people are arranged. If you are sitting in this stand there is little chance of moving about.

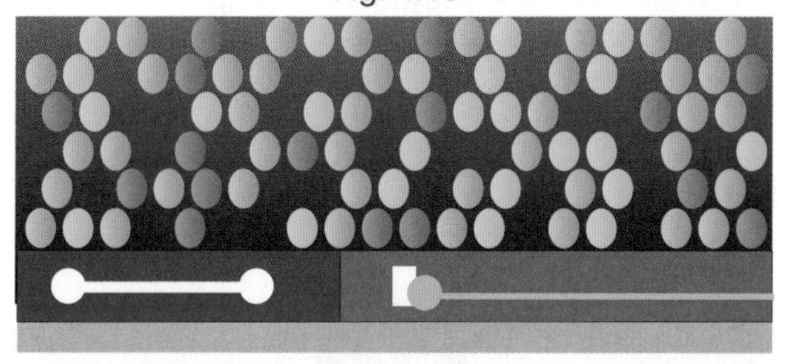

Fig. 4.11

In the second picture there is a crowd on the terrace. There is no regular arrangement of the people on the terrace. Although there are still a lot of people, there are gaps, so moving around is possible but not easy.

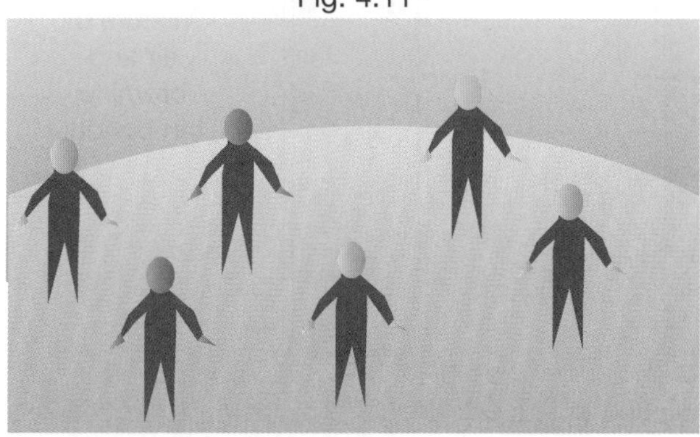

Fig. 4.12

In the third picture there are very few people on the grassy bank. There is no pattern in the way people are arranged on the bank. Because there are few people, it is easy to move about quickly.

These three pictures give a very good idea of how very tiny particles are arranged and move in solids, liquids and gases. These particles are so small they cannot be seen, even with a powerful microscope.

It is difficult in a diagram to appreciate the different movement of the particles in solids, liquids and gases.

These three diagrams show the arrangement of particles in solids, liquids and gases.

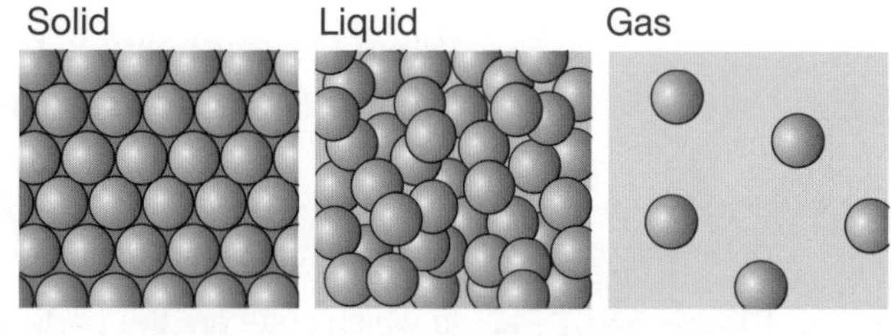

Fig. 4.13

The table compares the arrangement and movement of particles in solids, liquids and gases.

State	Arrangement of particles	Movement of particles
Solid	Particles closely packed together	Little movement – only vibrations
Liquid	Particles close together but not regularly arranged	Particles have some movement
Gas	Particles widely spaced	Particles moving rapidly in all directions

Using a particle model to explain observations

Explaining gas pressure

The particles in a gas are moving rapidly. There is no pattern to the movement. It is said to be random.

The particles hit the walls of the container. The more times the particles hit the walls, the higher the gas pressure.

If the temperature is raised, the particles move faster so there are more collisions with the walls of the container. As a result the pressure increases.

The gas is compressed into a smaller volume without changing temperature. Again, there will be more collisions with the walls. So there is an increase in pressure.

Explaining diffusion

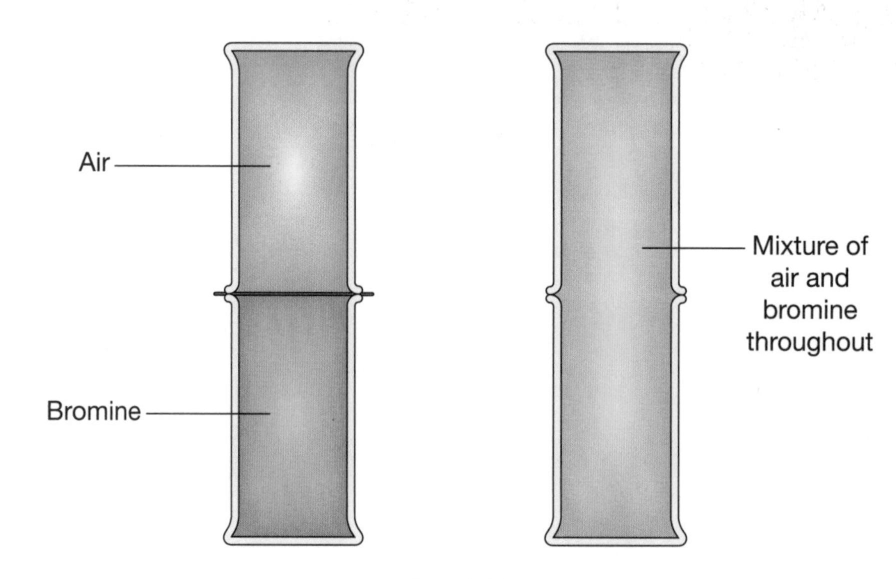

Fig. 4.14

In the diagram there are two gas jars separated by a piece of card. One gas jar contains brown bromine gas and the other contains air.

When the cardboard is removed, particles move from one jar to the other. Eventually the two gas jars contain the same mixture of bromine and air.

Progress Check

1. In which state – solid, liquid or gas – are the particles most widely spaced?
2. In which state are the particles moving fastest?
3. In which state are the particles regularly arranged?

1. Gas 2. Gas 3. Solid

4.4 Solutions

After studying this topic you should be able to:

- **classify solids as soluble or insoluble**
- **explain the meaning of the term saturated solution**
- **describe how mixtures can be separated by distillation and chromatography**
- **use the particle model to explain what happens when a solid dissolves in water**

Making solutions

When salt is added to water, the salt disappears from view. The water is still there. The water tastes salty. The salt has dissolved and formed a salt **solution**. A substance that dissolves is said to be **soluble** and a substance that does not dissolve is said to be **insoluble**.

A saturated solution of salt contains 36 g of salt dissolved in 100 g of water at room temperature.

If more and more salt is added to water, a point is reached where no more salt will dissolve. The undissolved salt sinks to the bottom. The solution formed contains the maximum mass of dissolved salt in a given mass of water at room temperature. It is called a **saturated solution**.

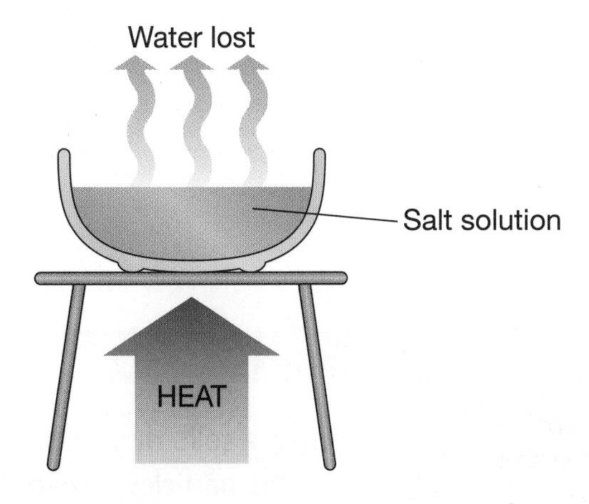

Fig. 4.15 Recovering salt from a salt solution.

Salt can be recovered from a salt solution by **evaporation**. When the solution is heated the water boils and turns to steam.

Separating mixtures

Water can be recovered from a salt solution by **distillation**. The salt solution is heated. The water boils and escapes from the flask. The steam escapes into the condenser. Here the steam **condenses** and forms liquid water. This is collected in the receiver.

Distillation involves evaporation followed by condensation.

Distillation can be used to remove a liquid from solid impurities.

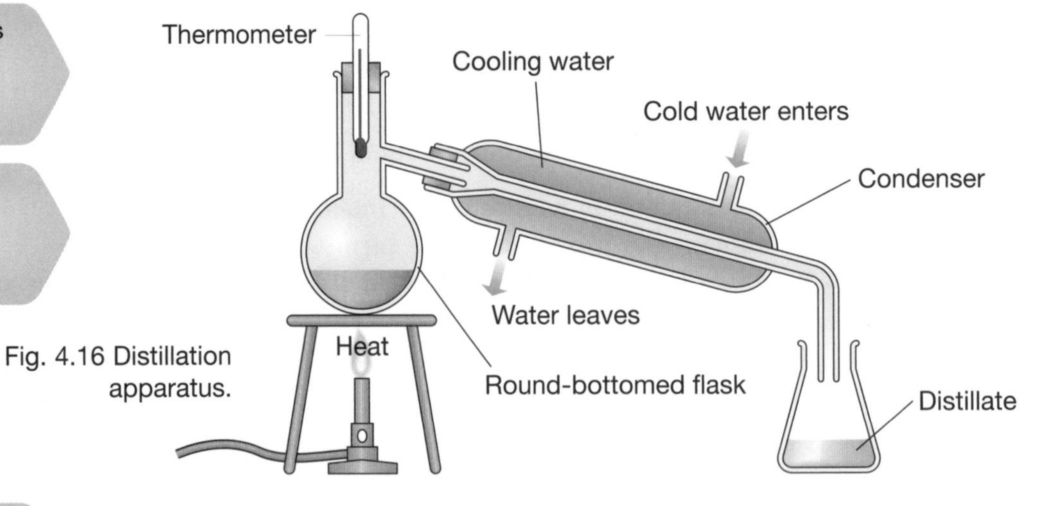

Fig. 4.16 Distillation apparatus.

Chromatography is often used to separate coloured substances, for example to separate the dyes in ink in a felt pen.

Mixtures of substances dissolved in water can be separated by **chromatography**. A drop of the solution containing a mixture of dissolved substances is placed at the bottom of a piece of filter paper. This is dipped into water and left. Water rises up the filter paper and the different dissolved substances separate at different rates. Each different substance forms a spot on the filter paper.

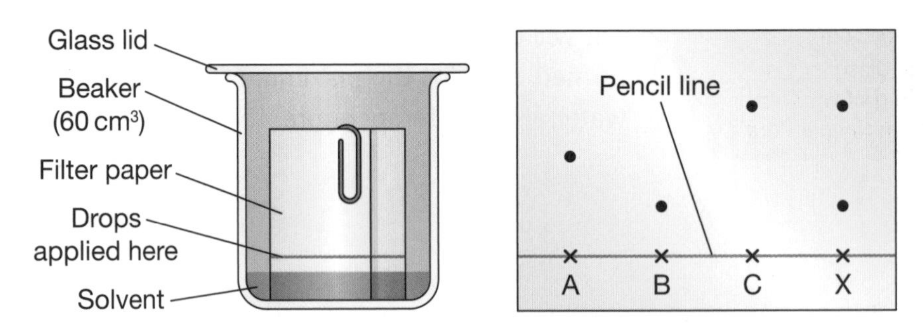

Fig. 4.17 Chromatography.

Particle model to explain dissolving

When a solid dissolves in water, the mass of the solution is the same as the combined mass of the solid and water.

When a solid dissolves, the regular arrangement of particles breaks down and the particles spread throughout the solution. They fill spaces between water particles. Solids dissolve faster in hot water because the particles have more energy and move faster.

Progress Check

Here is a list of substances.

chalk **salt** **sand** **sugar**

1 Which substances in the list are soluble in water and which are insoluble?

2 A scientist wants to separate the dyes in a fruit squash. Which method should be used?

3 In some countries drinking water is made from sea water. Which method should be used?

1. Soluble in water – sugar and salt. Insoluble in water – chalk and sand.
2. Chromatography 3. Distillation

Practice test questions

The following questions test levels 3-6

Shareen is making an indicator to test acids and alkalis.
She is using pink rose petals.

Here are the things she is doing. They are in the wrong order.

A Add the mixture plus a little ethanol to a boiling tube.
B Filter off the remains of the rose petals.
C Add a little ethanol and grind the petals using a pestle and mortar.
D Cut up the petals into small pieces.
E Heat the boiling tube and contents in a beaker of boiling water.

(a) Put these in the correct order. The first one has been done for you. **[3]**

D				

(b) Why is the boiling tube heated using a beaker of boiling water rather than a Bunsen burner flame? **[1]**

....

The table shows the colour the pink rose petal solution turns when added to sulphuric acid, water and sodium hydroxide solution.

	Sulphuric acid	**Water**	**Sodium hydroxide solution**
Colour of rose petal solution	Pink	Pink	Green

(c) What colour would pink rose petal solution turn in solutions of the following pH values? **[3]**

(i) pH 4 ....
(ii) pH 7 ....
(iii) pH 11 ....

(d) Why would this solution be of no use when trying to find out which of two solutions is an acid and which is a neutral solution? **[1]**

....

The following questions test levels 4-7

Nigel wants to make some copper(II) oxide from copper(II) carbonate.
He finds out two ways of doing this.

First way
Heat copper(II) carbonate in a test tube until it turns from a green powder to a black powder.

Second way
Add copper(II) carbonate to dilute hydrochloric acid. The mixture fizzes and a green solution is formed. Sodium hydroxide is added to the green solution and the test tube heated until a black solid is formed. The black solid is removed by filtering and dried.

(a) In the first way a gas is given off that turns limewater milky.

Write down the name of the gas produced. .... **[1]**

(b) **(i)** Finish the word equation for the reaction between copper(II) carbonate and hydrochloric acid. **[2]**

Copper(II) carbonate + hydrochloric acid → ________ + ________ + ________

(ii) Is the mass of the products greater than, less than or the same as the mass of the reactants? Explain your answer. **[2]**

....

....

(c) Using the first way, Nigel produces 8.0 g of copper(II) oxide from 12.2 g of copper(II) carbonate. What mass of copper(II) oxide would be produced from 12.2 g of copper(II) carbonate using the second way? **[1]**

....

Year 8

5 Chapter Five

The topics covered in this chapter are:

- **Atoms and elements**
- **Compounds and mixtures**
- **Rocks and weathering**
- **The rock cycle**

5.1 Atoms and elements

After studying this topic you should be able to:

- **recognise that there are about 100 known elements**
- **recall the names of some elements and their symbols**
- **identify elements that do not fit in as metals or non-metals**

Elements

You will know about a large number of different materials. Materials are all made up from basic building materials called **elements**. Sometimes a material is a single element, such as gold or sulphur. Sometimes it is made up from two or more elements.

Key Point

Elements are the basic building blocks from which all materials are made. There are about 100 known elements.

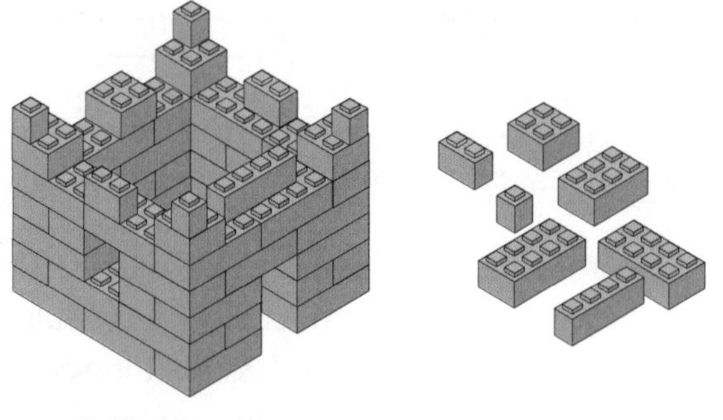

Fig. 5.1 A model made of bricks.

The picture shows a model made of bricks. You can see that this model is made up of a number of different types of brick. Using the same bricks we could make a number of other models.

Polyester and wood are two materials with widely different properties and uses.

An element is broken down to the smallest part that can exist. This is called an atom. Iron is made up of iron atoms.

Both polyester and wood are made of the same three elements – carbon, hydrogen and oxygen. The elements are put together in different ways just like the bricks in different models.

Scientists use shorthand to represent the elements. Each element is given a **symbol**.

The table gives the symbols of some common elements.

Element	Symbol	Element	Symbol
Aluminium	Al	Bromine	Br
Calcium	Ca	Carbon	C
Copper	Cu	Chlorine	Cl
Iron	Fe	Fluorine	F
Lead	Pb	Helium	He
Lithium	Li	Hydrogen	H
Magnesium	Mg	Iodine	I
Potassium	K	Nitrogen	N
Silver	Ag	Oxygen	O
Sodium	Na	Phosphorus	P
Zinc	Zn	Sulphur	S

The same symbols are used by scientists throughout the world.

These elements can be grouped in various ways.

Most of the known elements are **metals**. There are fewer **non-metals**. In the table above the elements in the left-hand column are metals and the elements in the right-hand column are non-metals.

The table gives information about some elements.

Element	Symbol	State at 20°C	Appearance	Conductor of electricity	Magnetic or not
iron	Fe	solid	shiny grey	conductor	magnetic
sulphur	S	solid	dull yellow	non-conductor	not magnetic
bromine	Br	liquid	dull red	non-conductor	not magnetic
copper	Cu	solid	shiny brown	conductor	not magnetic
nickel	Ni	solid	shiny grey	conductor	magnetic
selenium	Se	solid	shiny grey	semi-conductor	not magnetic

Looking at the data in the table, iron, copper and nickel are metals. They are shiny and metallic in appearance and they conduct electricity. Sulphur and bromine are non-metals. Selenium is difficult to classify. It looks metallic but does not conduct electricity like a metal. Unlike metals it breaks easily.

Classifying elements as metals or non-metals is not clear-cut. There are some elements with properties between metal and non-metal.

The elements are shown in the **periodic table**. The red line divides metals on the left-hand side from non-metals on the right.

Period	Group 1	Group 2	*d*-block transition elements										Group 3	Group 4	Group 5	Group 6	Group 7	Group 0
1						1 H Hydrogen 1												4 He Helium 2
2	7 Li Lithium 3	9 Be Beryllium 4											11 B Boron 5	12 C Carbon 6	14 N Nitrogen 7	16 O Oxygen 8	19 F Fluorine 9	20 Ne Neon 10
3	23 Na Sodium 11	24 Mg Magnesium 12											27 Al Aluminium 13	28 Si Silicon 14	31 P Phosphorus 15	32 S Sulphur 16	35.5 Cl Chlorine 17	40 Ar Argon 18
4	39 K Potassium 19	40 Ca Calcium 20	45 Sc Scandium 21	48 Ti Titanium 22	51 V Vanadium 23	52 Cr Chromium 24	55 Mn Manganese 25	56 Fe Iron 26	59 Co Cobalt 27	59 Ni Nickel 28	64 Cu Copper 29	65 Zn Zinc 30	70 Ga Gallium 31	73 Ge Germanium 32	75 As Arsenic 33	79 Se Selenium 34	80 Br Bromine 35	84 Kr Krypton 36
5	85.5 Rb Rubidium 37	88 Sr Strontium 38	89 Y Yttrium 39	91 Zr Zirconium 40	93 Nb Niobium 41	96 Mo Molybdenum 42	99 Tc Technetium 43	101 Ru Ruthenium 44	103 Rh Rhodium 45	106 Pd Palladium 46	108 Ag Silver 47	112 Cd Cadmium 48	115 In Indium 49	119 Sn Tin 50	122 Sb Antimony 51	128 Te Tellurium 52	127 I Iodine 53	131 Xe Xenon 54
6	133 Cs Caesium 55	137 Ba Barium 56	139 La Lanthanum 57	178.5 Hf Hafnium 72	181 Ta Tantalum 73	184 W Tungsten 74	186 Re Rhenium 75	190 Os Osmium 76	192 Ir Iridium 77	195 Pt Platinum 78	197 Au Gold 79	201 Hg Mercury 80	204 Tl Thallium 81	207 Pb Lead 82	209 Bi Bismuth 83	210 Po Polonium 84	210 At Astatine 85	222 Rn Radon 86
7	223 Fr Francium 87	226 Ra Radium 88	227 Ac Actinium 89	– Db Dubnium 104	– Jl Joliotium 105	– Rf Rutherfordium 106	– Bh Bohrium 107	– Hn Hahnium 108	– Mt Meitnerium 109									

key

atomic mass
symbol
name
atomic number

***f*-block**

139 La Lanthanum 57	140 Ce Cerium 58	141 Pr Praseodymium 59	144 Nd Neodymium 60	147 Pm Promethium 61	150 Sm Samarium 62	152 Eu Europium 63	157 Gd Gadolinium 64	159 Tb Terbium 65	162.5 Dy Dysprosium 66	165 Ho Holmium 67	167 Er Erbium 68	169 Tm Thulium 69	173 Yb Ytterbium 70	175 Lu Lutetium 71
227 Ac Actinium 89	232 Th Thorium 90	231 Pa Protactinium 91	238 U Uranium 92	237 Np Neptunium 93	242 Pu Plutonium 94	243 Am Americium 95	247 Cm Curium 96	247 Bk Berkelium 97	251 Cf Californium 98	254 Es Einsteinium 99	253 Fm Fermium 100	256 Md Mendelevium 101	254 No Nobelium 102	257 Lr Lawrencium 103

Fig. 5.2 The periodic table.

Elements can also be grouped as solids, liquids and gases.
Room temperature is 20°C.
An element with a melting point above 20°C is solid at room temperature.
An element with a boiling point below 20°C is gas at room temperature.
An element with a melting point below 20°C and a boiling point above 20°C is a liquid at room temperature.

Progress Check

1 The table gives the melting and boiling points of five elements. They are labelled A–E (these are not chemical symbols).

Is each element a solid, liquid or a gas at room temperature (20°C)?

Element	Melting point in °C	Boiling point in °C
A	-218	-183
B	1410	2360
C	-39	357
D	420	970
E	-7	58

2 Elements ending in -ium are usually metals.

Pick out the element in the list that is not a metal. (Hint: look at the periodic table.)

Aluminium **Barium** **Calcium** **Helium** **Sodium**

1. A – gas; B – solid; C – liquid; D – solid; E – liquid. 2. Helium

5.2 Compounds and mixtures

After studying this topic you should be able to:

- **recall that compounds are made when elements are joined**
- **distinguish symbols for elements and formulae for compounds**
- **distinguish elements, mixtures and compounds in terms of the particles they contain**
- **name and describe some common mixtures**

Elements combining to form compounds

You probably know that water is written as H_2O. This is called a formula and it shows us that two hydrogen atoms combine with one oxygen atom.

Certain mixtures of elements combine together to form compounds.

For example, hydrogen and oxygen are colourless gases. A mixture of hydrogen and oxygen explodes to form water (hydrogen oxide). The hydrogen and oxygen atoms join together to form liquid water.

Iron(II) sulphide is a compound formed when the elements iron and sulphur combine. The diagram shows what is happening when this reaction takes place.

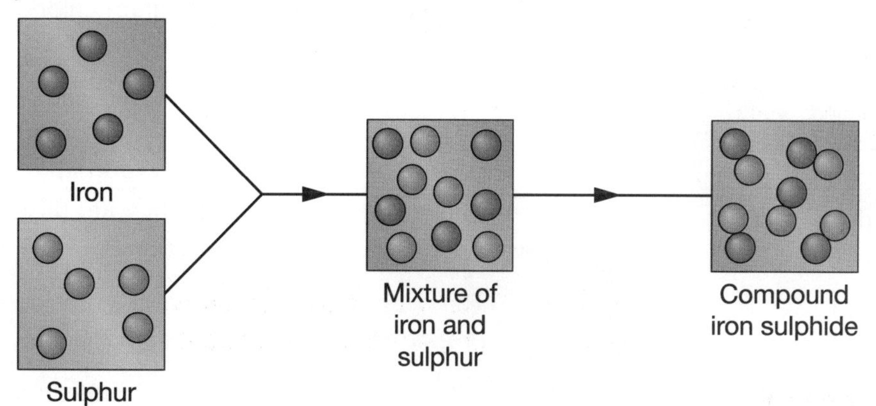

Fig. 5.3 The formation of iron(II) sulphide compound.

Compounds of two elements, e.g. sodium chloride and magnesium oxide, have names ending in -ide. Compounds ending in -ate contain three elements and one of them is oxygen.

The mixture of iron and sulphur can be separated using a magnet. Iron and sulphur cannot be separated from the compound iron sulphide. When the reaction takes place and a new substance is formed, one **atom** of iron joins with one **atom** of sulphur to form one **molecule** of iron(II) sulphide. That is why the **formula** of iron(II) sulphide is written as FeS.

The table gives the names and formulae of some common compounds. It also gives the numbers of atoms of the different elements present in the formula.

Name	Formula	Number of atoms of different elements present
Sodium chloride	NaCl	1 atom of sodium and 1 atom of chlorine
Magnesium oxide	MgO	1 atom of magnesium and 1 atom of oxygen
Calcium carbonate	$CaCO_3$	1 atom of calcium, 1 atom of carbon and three atoms of oxygen
Copper sulphate	$CuSO_4$	1 atom of copper, 1 atom of sulphur and 4 oxygen atoms
Sodium nitrate	$NaNO_3$	1 atom of sodium, 1 atom of nitrogen and 3 atoms of oxygen

Common mixtures

What is a pure substance? A pure substance is a single substance, element or compound. It contains no **impurities**. Many everyday substances are not pure elements or compounds but are **mixtures**.

Some common mixtures are:

Air

The composition of air varies. If a lot of people are in a crowded space, e.g. in an aeroplane, the percentage of oxygen decreases and carbon dioxide increases. The air has to be constantly changed by ventilation.

A mixture of gases containing about 80% nitrogen and 20% oxygen with small quantities of other gases. Its composition can be varied.

Air can be separated by fractional distillation. This involves cooling air until it is liquid (about -200°C) and allowing it to warm up. Different gases boil off at different temperatures.

Some solids are present in very small amounts in sea water. It is estimated that one cubic mile of sea water contains gold compounds worth £250 000 000. But it is not economic to extract them.

Sea water

Sea water consists of water with dissolved solids. These include a wide range of dissolved solids including sodium chloride (salt). The water can be removed by distillation.

Petrol

Petrol is a mixture of compounds of carbon and hydrogen. Since the compounds have similar properties, they are difficult to separate.

Key Point

The method used to separate a mixture depends upon the properties of the different components of the mixture.

There are many other examples of mixtures. Many rocks are mixtures of different minerals.

The table gives the differences between a pure substance (element or compound) and a mixture.

Pure substance	Mixture
Contains a single element or elements joined in fixed proportions.	Contains elements or compounds in proportions that are not fixed.
Difficult or impossible to separate elements in a compound.	Elements can be separated.
Properties of a compound are different from the elements combined together.	Properties of the mixture are the same as the properties of the things making it up.
Energy is often given out when a compound is formed.	No energy change when a mixture is formed.
Melts and boils at a fixed temperature.	Melts at a lower temperature than the pure substance and over a range of temperatures. Boils at a higher temperature than the pure substance.

Progress Check

1 The formula of aluminium sulphate is $Al_2(SO_4)_3$.

(a) How many elements are there in aluminium sulphate?

(b) How many atoms are there in one molecule of aluminium sulphate?

2 Look at the five diagrams.

1 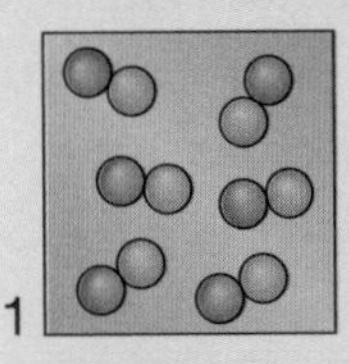 2 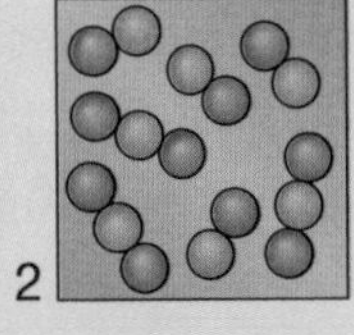 3 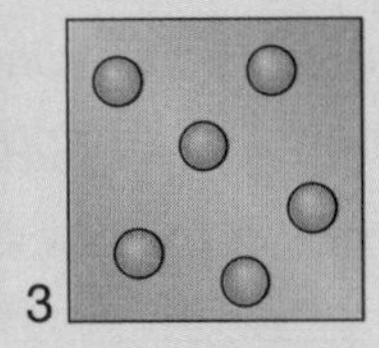 4 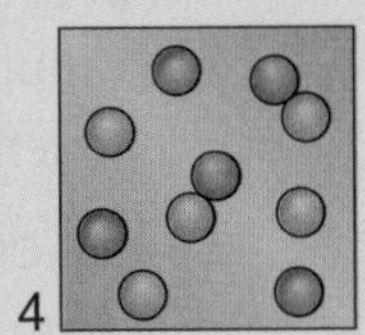 5 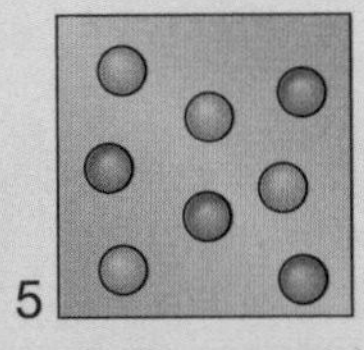

Which diagram represents

(a) an element?

(b) a pure compound?

(c) a mixture of elements?

(d) a mixture of compounds?

(e) a mixture of elements and compounds?

1. (a) Three (aluminium, sulphur and oxygen) (b) 17 2. (a) 3 (b) 1 (c) 5 (d) 2 (e) 4

5.3 Rocks and weathering

After studying this topic you should be able to:

- **describe rock specimens in terms of texture and relate this to properties such as porosity**
- **describe physical and chemical processes by which rocks are weathered**
- **describe how fragments of rock are transported and deposited as layers of sediment**

How old are the rocks of the Earth?

When thinking of the rocks of the Earth it is difficult to understand the lengths of time involved. It has taken 4500 million years for the Earth to be at the stage it has reached today. You get a better idea if we use a time-chart of the history of the Earth. If we take a year to represent the life of the Earth, each day represents 12.3 million years.

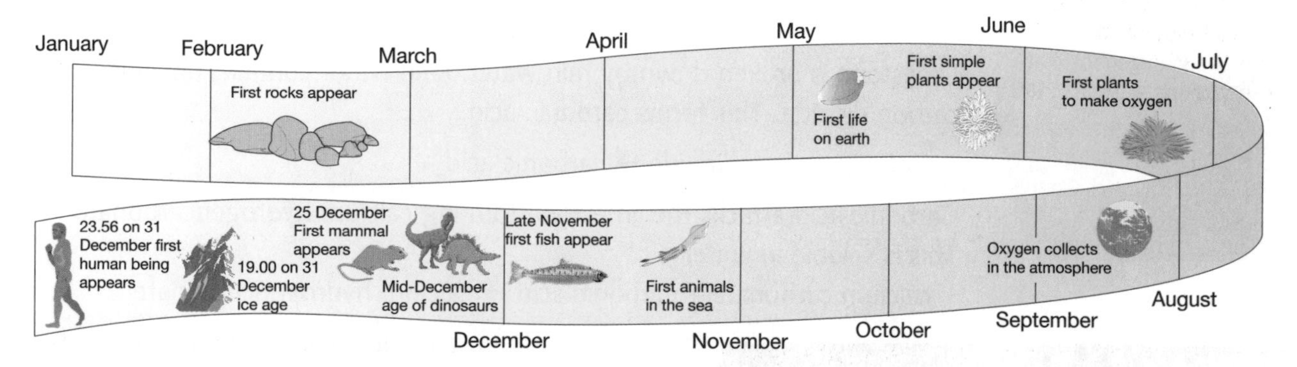

Fig. 5.4 The history of the Earth represented as one year.

What are rocks made from?

Rocks are made from a wide range of chemical compounds called **minerals**.

When you look at a rock carefully you might see different bits that fit together. Sometimes these bits are small **crystals** and sometimes they are small **grains**.

If they are very closely packed together then water does not pass through the rock. If there are gaps between these, water can pass through. These rocks are said to be **permeable**.

Weathering

Many buildings are made of rocks. Over a period of years these rocks are broken down by weathering. The photograph on the next page shows weathering of rocks at the Taj Mahal in India.

Fig. 5.5 Weathering of rocks at the Taj Mahal in India.

Weathering can take place in three main ways:

1 **Mechanical weathering**

During the day rocks heat up and expand. At night they cool and this cooling causes stresses within the rock. When this happens over and over again it breaks down the rock. If there are cracks in the rock, water gets into the cracks. When water freezes, ice forms and expands. This breaks down the rocks.

The terms 'weathering' and 'erosion' are often confused. Erosion involves the breaking down of rocks by movement of rivers, ice, sea and wind.

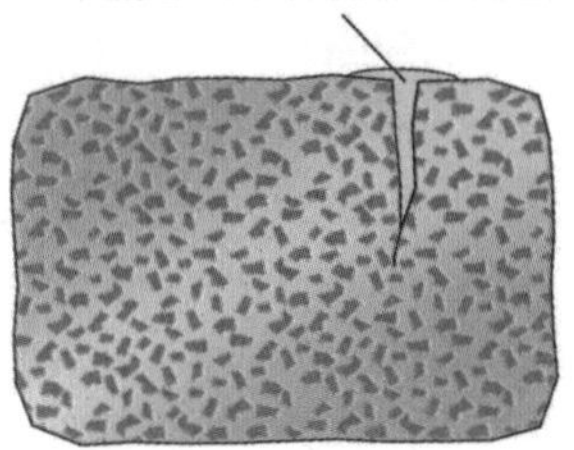

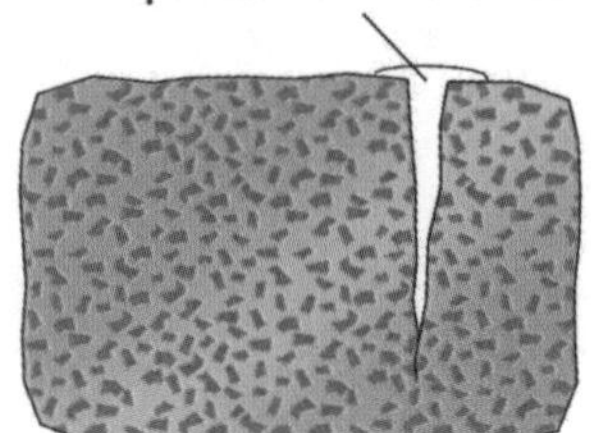

Fig. 5.6 Repeated freezing and thawing breaks down rocks.

This is sometimes called 'freeze-thaw'.

2 **Chemical weathering**

This is the breaking down of rocks by chemical reaction.

Limestone is broken down by rain water. Rain water contains dissolved carbon dioxide. This forms carbonic acid.

water + carbon dioxide → carbonic acid

Carbonic acid attacks the limestone forming calcium hydrogencarbonate. This is soluble in water.

calcium carbonate + carbonic acid → calcium hydrogencarbonate

You may see these word equations written as symbol equations.

$H_2O + CO_2 \rightarrow H_2CO_3$

$CaCO_3 + H_2CO_3 \rightarrow Ca(HCO_3)_2$

This weathering is speeded up when other acids are present in the atmosphere from acid rain.

3 **Biological weathering**

As plant roots grow they cause stresses on rocks and can cause them to break up.

Fig. 5.7 Biological weathering.

The rock fragments produced by these forms of weathering often get washed into rivers. As the fragments get carried along, they become more rounded, loosing sharp edges. As the speed of the river slows, the fragments are deposited on the riverbed. Heavy fragments drop first and fine fragments are carried further.

Conglomerate is a rock containing large fragments. This will be deposited close to where the river enters the sea. Shale is made from very fine particles and is formed away from the entry of the river into the sea.

Fig. 5.8 Conglomerate.

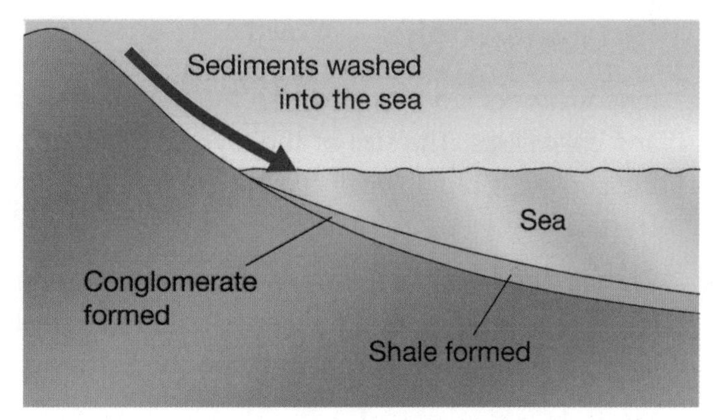

Fig. 5.9 Formation of conglomerate and shale.

Progress Check

1. What are the three types of weathering?
2. How does 'freeze-thaw' break up rocks?
3. The composition of conglomerate can be very varied. Suggest why?

1. Mechanical, chemical and biological. 2. Water in cracks freezes and expands, forces rock apart, thaws, process repeats. 3. Depends on which rocks are broken down.

5.4 The rock cycle

After studying this topic you should be able to:

- **describe how sediments become sedimentary rocks**
- **describe how metamorphic rocks are formed**
- **describe how igneous rocks can be formed by crystallising magma**
- **explain how different crystals of different sizes are formed**
- **explain how the rock cycle links the three types of rock**

Types of rock

There are three different types of rock in the Earth.

Much of Great Britain is a limestone area. Great Britain was once near the equator and covered with a tropical ocean. The limestone is the remains of the shells of sea creatures.

Sedimentary rocks

These rocks are formed when bits of existing rocks broken off by weathering and erosion settle as sediment. This is then compressed by other rocks and cemented together.

The rocks are not crystalline but are made up of grains. Sedimentary rocks may contain **fossils**. Chalk and limestone are sedimentary rocks.

Metamorphic rocks

Metamorphic rocks are usually much harder than sedimentary rocks.

Metamorphic rocks are formed when high temperatures and high pressures act on sedimentary rocks. These rocks may be non-crystalline or contain tiny crystals.

Metamorphic rocks may contain distorted crystals. Marble is a metamorphic rock made from limestone. Slate is a metamorphic rock made from action of high temperatures and high pressures on mud.

Igneous rocks cannot contain fossils.

Looking at a sample you will see crystals of different minerals – quartz is white, mica is shiny black and feldspar is brown.

Igneous rocks

Igneous rocks are formed when the hot magma inside the Earth is cooled and crystallises. The size of the crystals depends on the rate of cooling. If the cooling is rapid, small crystals are formed. Slow cooling produces larger crystals.

Granite and basalt are two igneous rocks. Granite has larger crystals as it crystallises slowly inside the Earth. Basalt has small crystals as the crystallisation takes place rapidly on the Earth's surface.

Rock cycle

Inside the Earth, existing rocks can be taken back into the magma and new rocks formed. This is summarised in the diagram.

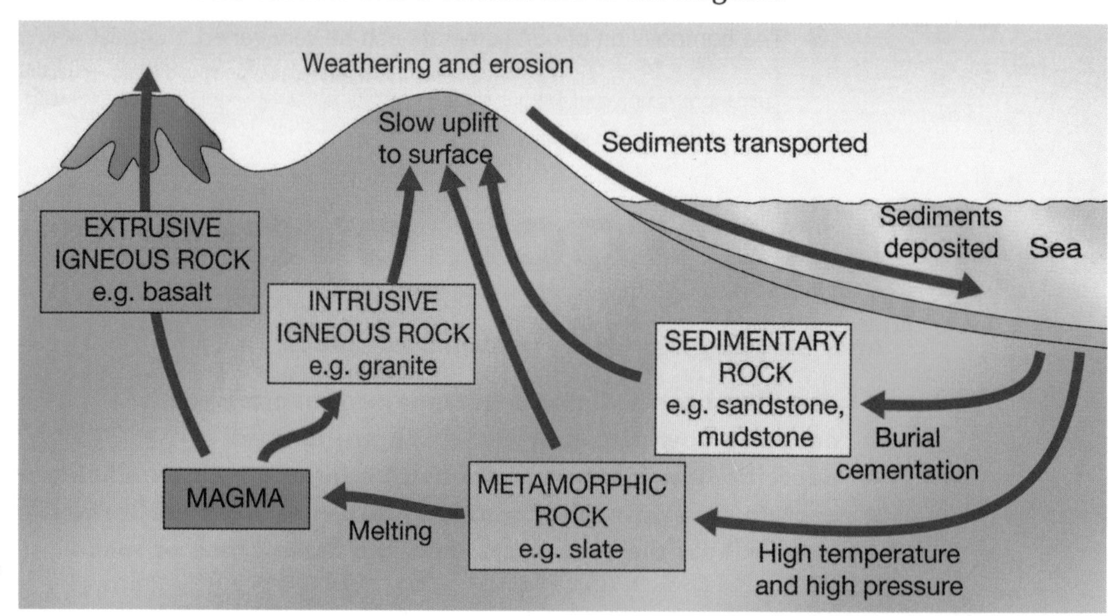

Fig. 5.10 The rock cycle.

Rocks are being broken down and new rocks are being formed all the time. When the magma crystallises, igneous rocks are produced. These are broken down by weathering and erosion to form sediments. Sediments are deposited and converted into sedimentary rocks. Sedimentary rocks can be converted into metamorphic rocks by high temperatures and high pressures. Rocks returning to the magma complete the cycle.

The rock cycle is driven by two energy processes. On the surface, processes are powered by the Sun's energy. Within the Earth, energy is provided by radioactive decay.

Progress Check

Which processes are used when

(a) sedimentary rocks are converted into metamorphic rocks?
(b) magma is converted into igneous rocks?
(c) sediments are converted into sedimentary rocks?
(d) igneous rocks are broken into sediments?

(a). High temperature and high pressure. (b). Cooling and crystallising. (c). Transported, settles, compressed (or compacted), cemented. (d). Weathering and erosion.

Practice test questions

The following questions test levels 3-6

The table gives some information about four substances.

Substance	Melting point in °C	Conductor of electricity	Appearance
Sulphur	114	No	Yellow solid
Paraffin wax	50–70	No	White solid
Mercury	-39	Yes	Silver liquid
Oxygen	-218	No	Colourless gas

(a) What information in the table suggests that mercury is a metal? **[2]**

....

....

(b) What information in the table suggests that paraffin wax is a mixture of substances but the other substances are pure? **[2]**

....

....

(c) About two hundred years ago the French scientist Lavoisier found that a red solid was formed when mercury and oxygen were heated together.

(i) Choose the word from the list that best describes the red solid. [1]

compound **element** **mixture**

....

(ii) Give a chemical name for the red solid. **[1]**

....

(d) Sulphur and oxygen react together to form sulphur dioxide, SO_2.

(i) Write a word equation for this reaction. **[2]**

....

(ii) How many atoms are there in a molecule of sulphur dioxide? **[1]**

....

The following questions test levels 4-7

Kim looks carefully at two rock samples labelled **A** and **B**.
She writes her observations in a table.

	Rock A	**Rock B**
Colour	Some white crystals and some shiny black ones	White crystals with some brown veins running through it
Size of crystals	Large	Small

(a) Why does she think that neither of these rocks is sedimentary? **[1]**

....

(b) What evidence is there in the table to suggest that Rock A is made up of more than one mineral? **[1]**

....

(c) What evidence is there to suggest that Rock A was crystallised slowly? **[1]**

....

(d) Rock B is a metamorphic rock formed from a sedimentary rock.

What conditions are needed to turn a sedimentary rock into a metamorphic rock? **[2]**

....

....

6 Chapter Six

The topics covered in this chapter are:

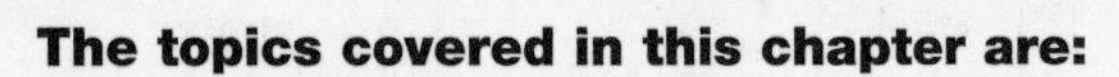

- **Reactions of metals and metal compounds**
- **Patterns of reactivity**
- **Environmental chemistry**
- **Using chemistry**

6.1 Reactions of metals and metal compounds

After studying this topic you should be able to:

- **recall some uses of metals**
- **describe how acids react with metals and metal compounds**
- **identify patterns in reactions and explain the use of these**
- **describe tests for hydrogen and carbon dioxide**
- **write symbol equations and make predictions about other reactions**

Uses of metals

It is hard to think of a world without metals. Different metals have different uses.

The picture shows metals in six different uses.

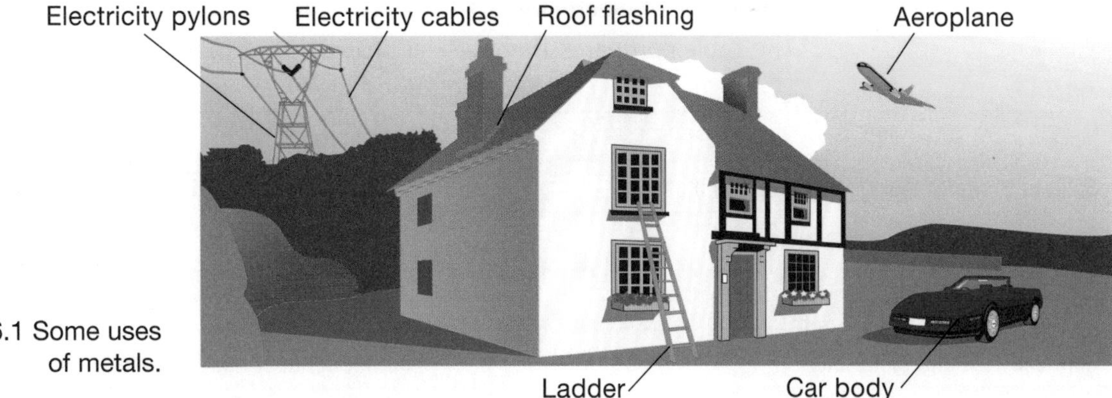

Fig. 6.1 Some uses of metals.

These metals have been chosen because they have different and useful properties.

The table gives a suitable metal for each use and some reasons why it is used.

Use	Metal	Reason for use
car body	steel (an alloy of iron)	strong; can be made into sheets
aeroplane	aluminium alloy	low density; strong
electricity cables	pure aluminium	good conductor of electricity; light
pylons	steel	strong
flashing on house roof	lead	unreactive; soft so that it can be shaped
ladder	aluminium alloy	low density; strong

Looking at these uses, there are different reasons why a particular metal is used. The following points are worth remembering:

- In many cases a mixture of metals, called an **alloy**, is used rather than a pure metal. Steel is an alloy of iron with a small amount of carbon added. Aluminium alloy is much stronger than pure aluminium.
- Metals are often used because they are good **conductors of electricity**. Pure aluminium is a better conductor of electricity than aluminium alloy. The fact that aluminium alloy is stronger is not relevant. The best conductor of electricity is silver but it is not used widely for conducting electricity because it is too expensive.
- Metals are also **good conductors of heat**. Metals are used to make radiators in houses.
- Unreactive metals, such as gold, platinum and silver, are used for jewellery. This is because they are hard and unreactive.

Metals are good conductors of heat and electricity. Non-metals are bad conductors of heat and electricity.

Graphite is a form of carbon. Unlike other non-metals, graphite is a good conductor of electricity.
The table compares the physical properties of metals and non-metals.

Metals	Non-metals
Solid at room temperature (exception mercury)	Solid, liquid or gas at room temperature
Shiny	Dull
High density	Low density
Conduct heat and electricity	Do not conduct heat and electricity
Can be beaten into thin sheets (malleable) or drawn into thin wires (ductile)	Easily broken (brittle)

Progress Check

The table gives some information about three elements, labelled A, B and C.

Use the information to decide whether each one is a metal or a non-metal.

A	B	C
Gas	Solid	Solid
Very low boiling point	Shiny	Dull, brittle
Non-conductor of electricity	Conductor of heat and electricity	Conductor of electricity

A – non-metal; B – metal; C – non-metal

Reactions of acids

With metals

Key Point

Dilute acids react with some metals to produce a salt and hydrogen gas.

For example:

magnesium + hydrochloric acid → magnesium chloride + hydrogen

$Mg(s) + 2HCl(aq) \rightarrow MgCl_2(aq) + H_2(g)$

zinc + sulphuric acid → zinc sulphate + hydrogen

$Zn(s) + H_2SO_4(aq) \rightarrow ZnSO_4(aq) + H_2(g)$

These equations have state symbols added in brackets. The symbols mean (s) solid, (aq) solution with water as the solvent, (g) gas and (l) liquid.

We can test for hydrogen by putting a **lighted splint** into the test tube. The hydrogen burns with a **squeaky pop**.

After the test there are droplets of colourless liquid in the test tube that was dry at the start. This liquid is hydrogen oxide (water). It was formed when hydrogen burned.

hydrogen + oxygen → water

With metal carbonates

Key Point

Dilute acids react with carbonates to form a salt, water and carbon dioxide.

For example:

calcium carbonate + hydrochloric acid → calcium chloride + water + carbon dioxide

$CaCO_3(s) + 2HCl(aq) \rightarrow CaCl_2(aq) + H_2O(l) + CO_2(g)$

sodium carbonate + sulphuric acid → sodium sulphate + water + carbon dioxide

$Na_2CO_3(s) + H_2SO_4(aq) \rightarrow Na_2SO_4(aq) + H_2O(l) + CO_2(g)$

With metal oxides

Key Point

Dilute acids react with metal oxides to form a salt and water only.

Metal oxides are sometimes called bases.

For example:

copper(II) oxide + sulphuric acid → copper(II) sulphate + water

$CuO(s) + H_2SO_4(aq) \rightarrow CuSO_4(aq) + H_2O(l)$

zinc oxide + hydrochloric acid → zinc chloride + water

$ZnO(s) + 2HCl(aq) \rightarrow ZnCl_2(aq) + H_2O(l)$

The diagram shows how a sample of copper(II) sulphate can be made from copper(II) oxide.

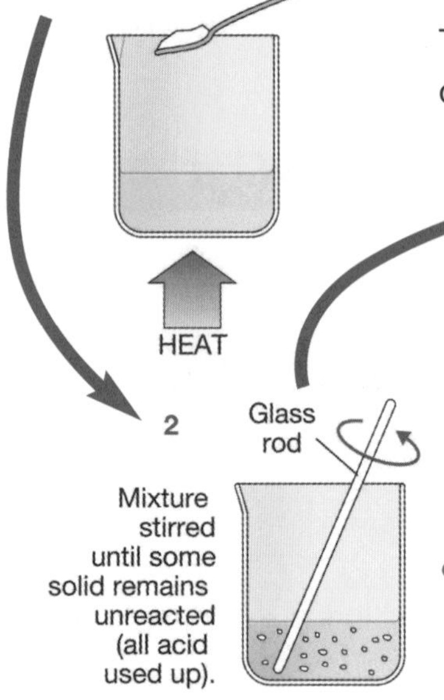

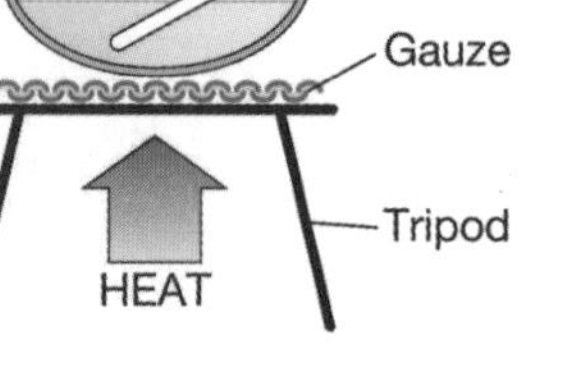

Fig. 6.2 Preparation of copper(II) sulphate.

- In 2 the frequent mistake is to write that the solution is saturated with copper(II) oxide. This assumes the copper(II) oxide is dissolving. It is reacting. The reaction stops when all the acid is used up.
- No bubbles are seen when copper(II) oxide and sulphuric acid react because no gas is produced.
- Filtering removes excess copper(II) oxide.
- Evaporating the solution to dryness would decompose the copper(II) sulphate crystals.

With metal hydroxides

Key Point

A metal hydroxide reacts with an acid to form a salt and water only.
Hydrochloric acid is used to make chlorides, sulphuric acid makes sulphates and nitric acid makes nitrates.

In Chapter Four you found out that acids and alkalis react in neutralisation reactions.

For example:

sodium hydroxide + hydrochloric acid → sodium chloride + water

$NaOH(aq) + HCl(aq) \rightarrow NaCl(aq) + H_2O(l)$

sodium hydroxide + sulphuric acid → sodium sulphate + water

$2NaOH(aq) + H_2SO_4(aq) \rightarrow Na_2SO_4(aq) + 2H_2O(l)$

sodium hydroxide + nitric acid → sodium nitrate + water

$NaOH(aq) + HNO_3(aq) \rightarrow NaNO_3(aq) + H_2O(l)$

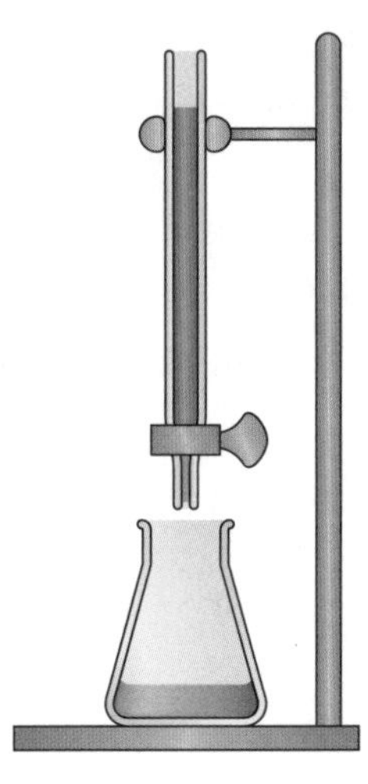

25 cm^3 of sodium hydroxide solution is added to the conical flask. A pH meter probe is put into the solution.
The pH reading is 13.

Hydrochloric acid is added drop by drop until the solution turns green. The solution is now neutral.

Fig. 6.3 Hydrochloric acid is added drop by drop.

The solution is evaporated to recover the salt, sodium chloride.

Progress Check

1 Complete the following word equations:

(a) sodium carbonate + hydrochloric acid → _______ + _______ + _______

(b) magnesium oxide + sulphuric acid → _______ + _______

2 Calcium reacts with dilute sulphuric acid and dilute hydrochloric acid.

(a) Predict the word equations and symbol equations for these reactions.

(b) Calcium sulphate is almost insoluble. What effect does this have on the reaction?

1. (a) sodium chloride + water + carbon dioxide (b) magnesium sulphate + water
2. (a) calcium + sulphuric acid → calcium sulphate + hydrogen
$Ca + H_2SO_4 \rightarrow CaSO_4 + H_2$
Calcium + hydrochloric acid → calcium chloride + hydrogen
$Ca + 2HCl \rightarrow CaCl_2 + H_2$
(b) Insoluble calcium sulphate forms a layer around the calcium carbonate preventing acid from attacking it. So little reaction seen.

Salts

Salts are compounds formed when an acid reacts with a metal, metal oxide, metal hydroxide or metal carbonate.

The table gives the names, formulae and uses of some salts.

Chemical name	Formula	Common name	Use
Sodium chloride	$NaCl$	Common salt	Flavouring and preserving food
Sodium carbonate	$Na_2CO_3.10H_2O$	Washing soda	Chemical raw material, softening water
Sodium sulphate	$Na_2SO_4.10H_2O$	Glauber's salts	Medicine
Potassium nitrate	KNO_3	Saltpetre	Making gunpowder
Magnesium sulphate	$MgSO_4.7H_2O$	Epsom salt	Laxative
Sodium stearate	Complicated		Soap

Progress Check

1. Calcium phosphate is a salt made by reacting calcium carbonate with an acid. Suggest the name of the acid.
2. Suggest the name of the acid used to make sodium carbonate. (Hint: It is the acid formed when carbon dioxide dissolves in rainwater.)
3. Write a word equation for the reaction producing potassium nitrate from potassium hydroxide.

1. phosphoric acid 2. carbonic acid
3. potassium hydroxide + nitric acid → potassium nitrate + water

6.2 Patterns of reactivity

After studying this topic you should be able to:

- **describe how metals react with water and oxygen**
- **use the reactivity series to make predictions about reactions of metals**
- **relate the uses of metals to their reactivity**

Reactions with water

Some metals react with cold water.

Key Point

When a metal reacts with water, hydrogen gas is produced.

Potassium and sodium react with cold water. When a piece of sodium is placed on the surface of water, the following observations can be made:

- It floats on the surface.
- It fizzes (producing a colourless gas).
- It produces a colourless solution. Universal indicator turns purple showing that an alkali (sodium hydroxide) is produced.

A common mistake here is to give the product of the reaction as sodium oxide, Na_2O, and not sodium hydroxide.

The equation for the reaction is:

sodium + water → sodium hydroxide + hydrogen

$2Na(s) + 2H_2O(l) \rightarrow 2NaOH(aq) + H_2(g)$

Calcium reacts very slowly with cold water. Again an alkali (calcium hydroxide) and hydrogen are produced.

calcium + water → calcium hydroxide + hydrogen

$Ca(s) + 2H_2O(l) \rightarrow Ca(OH)_2(aq) + H_2(g)$

Reactions with air or oxygen

Key Point

Some metals burn in air or oxygen to produce oxides – compounds of the metal with oxygen.

Sodium burns in oxygen to form solid sodium oxide.

sodium + oxygen → sodium oxide
$4Na(s) + O_2(g) \rightarrow 2Na_2O(s)$

Magnesium burns in oxygen to form solid magnesium oxide.

magnesium + oxygen → magnesium oxide
$2Mg(s) + O_2(g) \rightarrow 2MgO(s)$

Reactivity series

The reactions of metals with air or oxygen, with water and with dilute acids can be used to put metals in order of reactivity.

The information is summarised in the table.

Metal	Heated in oxygen	Reaction with water or steam	Reaction with acids
Potassium	↑	↑	Violent reaction
Sodium		Reacts with cold water	↓
Calcium	Burns to form an oxide	↓	↑
Magnesium		↑	
Aluminium		Reacts with steam	Reacts with dilute acids
Zinc	↓	↓	
Iron	Reacts slowly	Partial reaction with steam	↓
Copper	Partial reaction	No reaction	No reaction

Other metals can be included in this list.

The list of metals in order of decreasing reactivity is called the **reactivity series.**
The reactivity series often seen is shown below:

You do not have to try to remember this list of metals. It will be given to you when you need it.

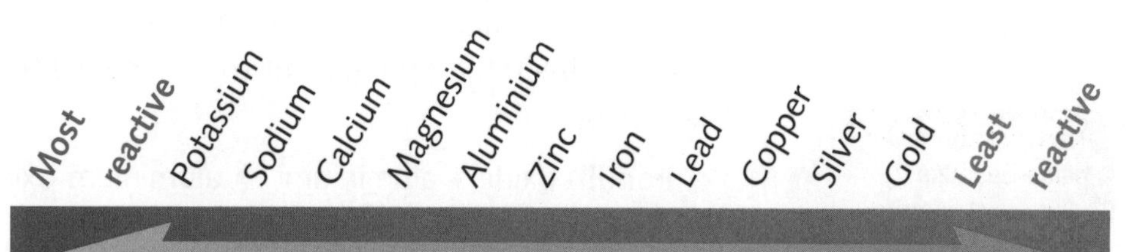

The use for a particular purpose may depend on its reactivity. Lead is used for flashings on a roof. One reason for this is its lack of reactivity.

Predicting reactions

The reactivity series is useful for predicting likely chemical reactions. For example, if iron filings are added to blue copper(II) sulphate solution, a brown solid is formed and the solution turns colourless. The iron and copper change places and free copper is formed.

Fig. 6.4

The situation is similar to the one in the cartoon where the more powerful robber is stealing money from the old man. The robber represents iron, the old man copper and the money sulphate.

The reaction can be summarised by the equation:

copper(II) sulphate + iron → iron(II) sulphate + copper

$CuSO_4(aq) + Fe(s) \rightarrow FeSO_4(aq) + Cu$

This reaction is called a displacement reaction. It takes place because iron is more reactive (higher in the reactivity series) than copper.

Fig. 6.5

In the second cartoon the old man is thinking about stealing money from the security guard. He cannot do it because the security guard is too powerful. This is similar to a situation where silver powder is added to copper(II) sulphate solution. The reaction does not take place because copper is more reactive than silver.

A useful displacement reaction is the Thermit reaction. This is used to weld together lengths of railway track on site.

Long lengths of track make the journey more comfortable but are difficult to transport. A mixture of aluminium powder and iron(III) oxide powder is placed between the two ends of track. The mixture is set alight. A reaction takes place:

This reaction takes place because aluminium is more reactive than iron.

iron(III) oxide + aluminium → aluminium oxide + iron

$Fe_2O_3(s) + 2Al(s) \rightarrow Al_2O_3(s) + 2Fe(l)$

The heat released is enough to melt the iron, which flows into the gap between the rails and welds them together.

Progress Check

Sara tested four metals with different metal salt solutions.
She put a tick (✓) if there was a reaction and a cross (x) if there was no reaction. Finish her table.

Metal	Magnesium chloride solution	Copper(II) chloride solution	Iron(III) chloride solution	Zinc chloride solution
Magnesium	X	✓		
Copper	X	X		
Iron	X	✓		
zinc	X	✓		

Metal chloride solution	Magnesium chloride solution	Copper(II) chloride solution	Iron(III) chloride solution	Zinc
Magnesium	X	✓	✓	✓
Copper	X	X	X	X
Iron	X	✓	X	X
zinc	X	✓	✓	X

6.3 Environmental chemistry

After studying this topic you should be able to:

- **describe some of the consequences of acid rain**
- **identify why it is important to monitor air and water pollution**
- **describe a variety of environmental issues**

Acid rain

The pH value of rain water is about 5.5.

Rain water is naturally slightly acidic because carbon dioxide dissolves in water forming carbonic acid.

water + carbon dioxide → carbonic acid

However, the burning of fossil fuels, such as coal, produces sulphur dioxide. This dissolves in water forming sulphuric acid.

Vehicle exhausts produce oxides of nitrogen (sometimes written as NO_x). These dissolve in water to form nitric acid.

What are the consequences of acid rain? They include:

1. Damage to stonework on buildings. St Paul's cathedral and Westminster Abbey are just two buildings that show damage due to acid rain. A black skin first appears on the stone. This then blisters and cracks, causing the stone to be seriously disfigured.
2. Rivers and lakes can become more acidic. This kills wildlife including fish and otters. There are many lakes in Norway or Sweden that now have no life.
3. Forests are seriously damaged. Forests in Scandinavia and Germany especially are being damaged by acid rain. Trees are stunted, needles and leaves drop off and the trees die. It has been estimated that acid rain is costing the German forestry industry about £250 million each year.
4. Human life can be affected. Acid conditions can alter levels of copper, lead and aluminium in the body. These changes have been linked with diarrhoea in small babies and breathing disorders.
5. Damage to metalwork. Acid rain can speed up corrosion of metals. Wrought iron railings in city areas can show considerable damage.

Monitoring air and water pollution

Monitoring levels of pollution in air and water is important in detecting changes in pollutants over a period of time. In Great Britain about 50 years ago levels of air pollutants were much higher than they are today. In cities there were frequent serious fogs produced by soot ash and tar from the burning of coal in houses and factories. This fog was nicknamed smog and caused 4000 extra deaths in one winter alone in Great Britain.

The Clean Air Act (1956) set up clean air zones where coal could not be burned, only 'smokeless fuels'. As a result of this act and the careful monitoring of air pollution the situation in cities has improved.

There are websites where you can get data about air pollution.

In Santiago, Chile, pollution caused by motor vehicles is controlled day to day by monitoring the levels of pollution and then banning vehicles with certain digits in their number plates for the next day.

Fig. 6.6 London before 1956.

Fig. 6.7 London today.

The problem today in cities is often due to pollution from cars and other motor vehicles. Monitoring levels of air pollutants can help planning for a healthier future in our cities.

Global warming

You have probably heard stories about the Earth warming up and some of the effects this might have. This is called **global warming** and is caused by the **greenhouse effect**.

In a greenhouse plants are kept warm. Short wavelength light energy from the Sun enters through the glass and warms up the inside of the greenhouse. The objects in the greenhouse radiate long wavelength infrared radiation but this cannot escape back out through the glass. The temperature then rises.

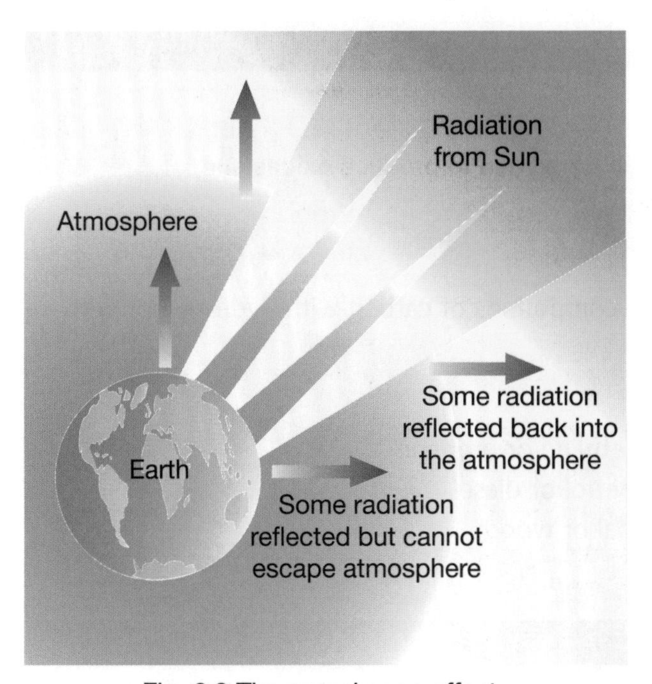

Fig. 6.8 The greenhouse effect.

The Earth acts in a similar way but, instead of glass, there are gases in the atmosphere (carbon dioxide and methane) that do the same job. They let the short wavelength energy in but do not let the long wavelength energy escape. Without the greenhouse effect the surface of the Earth would be too cold for life to exist. The problem of global warming comes about because the concentrations of carbon dioxide, in particular, are rising and the effect is getting greater and surface temperatures are rising.

Don't confuse the greenhouse effect with the destruction of the ozone layer.

Possible effects of global warming are:

- rising temperatures on the Earth's surface
- melting of glaciers and ice caps leading to rising sea levels
- rising ocean temperatures may affect plankton growth
- changes of climate in different parts of the world

Progress Check

Here is a list of gases. Use these gases to answer the following questions.

carbon dioxide methane oxygen ozone

1 Which two gases produce the greenhouse effect?
2 Which gas is produced when fossil fuels are burned?
3 Which gas screens out UV radiation from the Sun?

1. carbon dioxide and methane 2. carbon dioxide 3. ozone

6.4 Using chemistry

After studying this topic you should be able to:

- **recall the sorts of materials that are fuels and what is made when they burn**
- **explain the advantages and disadvantages of hydrogen as a fuel**
- **recall uses of chemical reactions in everyday situations**
- **recall and explain that mass is conserved in chemical reactions**

Energy from fuels

A fuel burns in air or oxygen to produce oxides and release energy.

Fuels are usually carbon or compounds of carbon with hydrogen or with hydrogen and oxygen.

Fuels can be:

gases – for example methane or propane
liquids – for example petrol or diesel
solids – for example coal or wood.

What makes a good fuel?

This is a difficult question to answer. It depends on where you are using the fuel. Petrol is good fuel in a car but it would be a bad fuel on a household fire.

Here is a list of things you might consider when choosing a suitable fuel.

1. Is the fuel a solid, liquid or gas?
2. Solid fuels are easy to store and liquids and gases need special storage tanks.
3. How easy is it to set alight?
4. Liquids and gases are generally easier to set alight and this might be important.
5. How long does it burn for?
6. What ash or other deposits are left behind?
7. Does the fuel burn cleanly and without smell?
8. What are the relative costs of the fuels?

When a carbon-based fuel is burned the products include **carbon dioxide**. This can increase global warming (see Chapter 6.3). Other products are formed, for example soot (carbon), carbon monoxide or water.

The products of burning carbon-based fuels depend on the amount of air or oxygen used. If carbon is burned in a plentiful supply of air, **carbon dioxide** is formed.

carbon + oxygen → carbon dioxide

$C(s) + O_2(g) \rightarrow CO_2(g)$

If carbon is burned in a short supply of air, carbon monoxide is formed.

carbon + oxygen → carbon monoxide

$2C(s) + O_2(g) \rightarrow 2CO(g)$

Carbon monoxide is a poisonous gas.

Every year in Great Britain about 50 people die of carbon monoxide poisoning. This is often due to gas appliances not being regularly serviced.

In a similar way fuels containing carbon and hydrogen or carbon, hydrogen and oxygen can produce different products in limited and plentiful supplies of air.

Plentiful supply of air:

methane + oxygen → carbon dioxide + water

Limited supply of air:

methane + oxygen → carbon monoxide + water

Progress Check

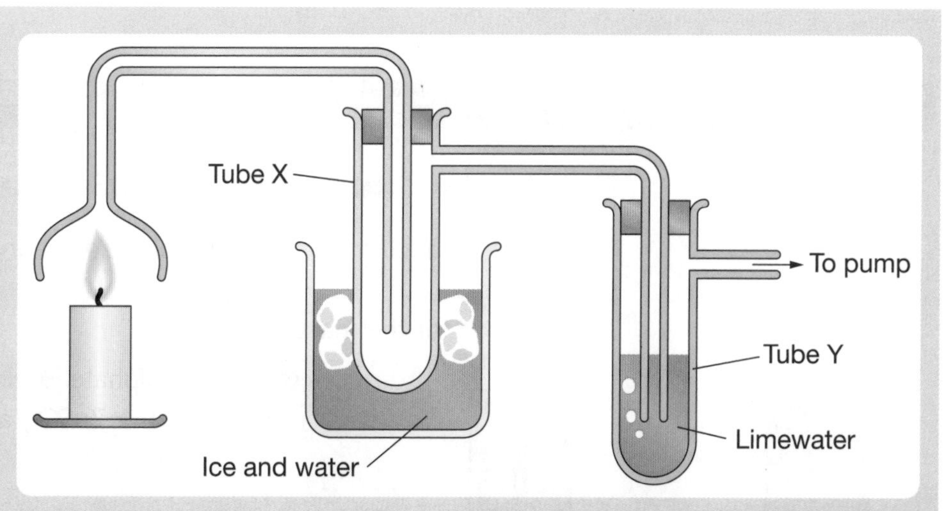

This apparatus can be used to show that carbon dioxide and water are formed when a candle burns. The fuel is candle wax.

A pump draws air though the apparatus. This air contains the products of the combustion of the candle. After a short while a colourless liquid collects in Tube X and the liquid in Tube Y turns white and cloudy.

1. How would you show that the liquid collected in Tube X is water?
2. What causes the liquid in Tube Y to go white?
3. Tim says that ordinary air contains water and carbon dioxide. How could you show him that it is the products of burning the candle that cause the changes in Tubes X and Y?
4. The experiment was repeated with a jet of burning hydrogen gas. How would the results be different?

1. Liquid boils at 100 °C. 2. Carbon dioxide 3. Run the experiment without the burning candle. There is little change in the two tubes after three minutes. With the candle in place the changes are seen very quickly. 4. Colourless liquid collected but limewater does not turn white.

Hydrogen as a fuel

Hydrogen has been suggested as a clean fuel. It is already used as a fuel for space rockets. Some buses have been converted to burn hydrogen gas and a Boeing passenger plane has been converted for a similar reason.

What are the advantages of hydrogen as fuel?

- It burns cleanly producing a large amount of energy.
- It does not form any carbon dioxide or carbon monoxide.
- The only product is water and that is not a pollutant.

What are the disadvantages of hydrogen?

- Most of the disadvantages come from storage. The hydrogen has to be stored as liquid. It has to be kept below –250 °C.
- A good source of hydrogen is required, as it is not naturally available.
- People are concerned that after an accident hydrogen might cause an explosion.
- Tests have shown that hydrogen is safer than paraffin. If the hydrogen tank splits, the hydrogen vaporises and the vapour quickly rises and escapes because hydrogen has such a low density. Paraffin remains a liquid and can catch alight.

Energy from other chemical reactions

Exothermic and endothermic reactions

When zinc is added to copper(II) sulphate solution, a displacement reaction takes place.
The equation for the reaction is:

zinc + copper(II) sulphate → zinc sulphate + copper
$Zn(s) + H_2SO_4(aq) \rightarrow ZnSO_4(aq) + Cu(s)$

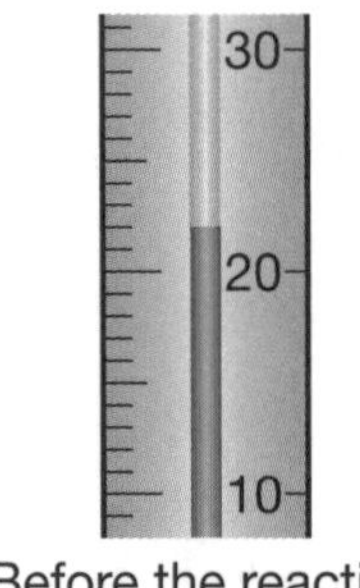

Before the reaction

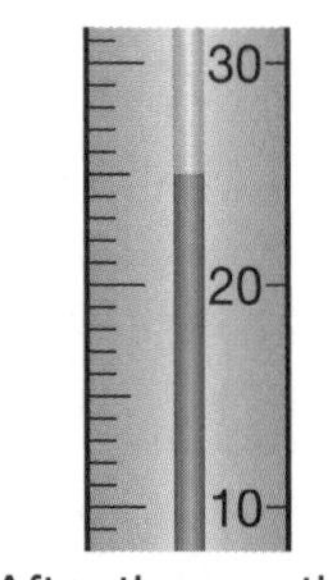

After the reaction

Fig. 6.9 The thermometers show the change in temperature due to the reaction.

The diagram shows part of the thermometer scale before and after the reaction. There is a 3 °C temperature rise. The energy has come from the reaction. Burning releases energy and so does this reaction. The energy is released as heat and causes the temperature of the solution to rise.

A reaction where energy is released causing a rise in temperature is called an exothermic reaction.

There are some reactions that take in energy from the surroundings and cause the temperature of the solution to fall. One example is the reaction of sodium hydrogencarbonate and citric acid. Reactions like this are called **endothermic** reactions.

There are many exothermic reactions. Scientists once thought that all reactions were exothermic.

Getting the energy from a reaction as electricity

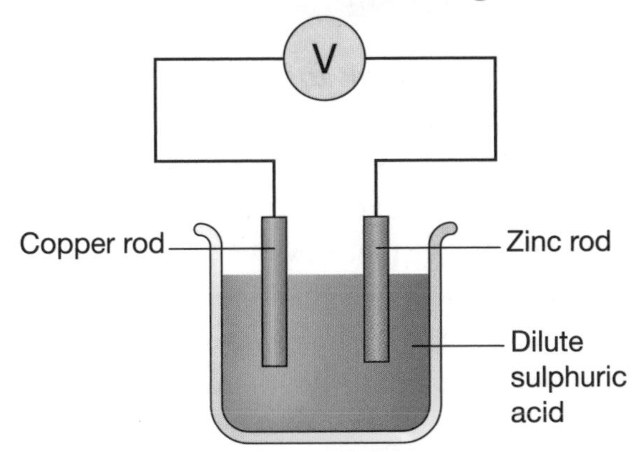

The diagram shows a rod of zinc and a rod of copper dipping into sulphuric acid solution. The two rods are connected to a voltmeter. The reading on the voltmeter shows that there is a potential difference between the two rods and a current could pass. This is a **simple cell**.

Fig. 6.10 A simple cell.

What happens to the voltage if the metals making the rods are changed?

1. There is no voltage if the two rods are made of the same metal.
2. The further apart the metals are in the reactivity series, the greater will be the voltage.

For example:

copper and zinc 1.1 V   copper and lead 0.47 V

Getting energy from rusting

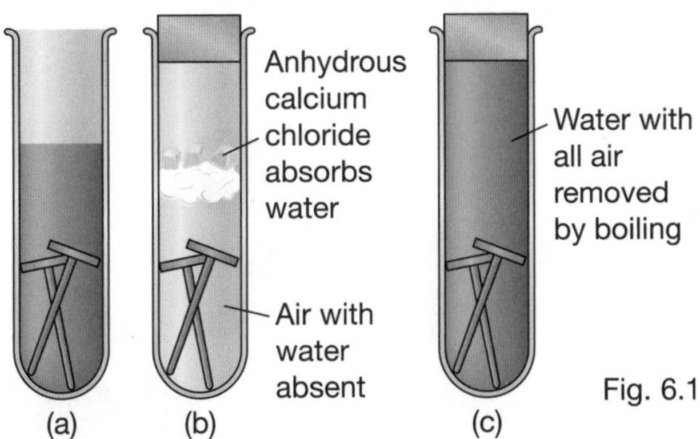

Iron and steel rust when left exposed to the air. If the test tubes are set up as in the diagram, rusting takes place in (a) where air and water are present. Rusting does not take place in (b) where air is present and not water. Rusting does not take place in (c) when water is present but air is not.

Fig. 6.11 Rusting of iron.

Rusting is similar to burning. Both processes involve reaction with oxygen and both are exothermic. This is not obvious in the case of rusting because it takes place over a long period of time.

This process can be used to produce a heater for heating food without using a stove. Iron filings rust in contact with air and water and good insulation prevents heat escaping. The food is therefore heated.

Progress Check

1. Which pair of metals gives the biggest voltage in a simple cell?

   silver and copper   **silver and magnesium**   **silver and iron**

2. The word equations for photosynthesis and respiration are:

   Carbon dioxide + water → glucose + oxygen
   Glucose + oxygen → carbon dioxide + water

   The respiration reaction is exothermic. What does this tell you about the photosynthesis reaction?

1. Silver and magnesium (further apart in the reactivity series) 2. Endothermic

Mass changes during a reaction

When a piece of magnesium burns in air, it burns to form a white solid. If this reaction had been known 200 years ago, scientists would have explained what was happening by a loss of a substance called **phlogiston** when magnesium burned. They would have explained that white solid does not burn because it does not contain phlogiston. However, if the magnesium and the white solid produced are weighed there is a gain in mass and not a loss. The gain of mass is due to the oxygen from the air when the magnesium combines with oxygen to form the white solid.

Key Point

The sum of the masses of the reactants is always equal to the masses of the products.

- If there is an apparent loss of mass it is because something (usually a gas) has been lost.
- If there is an apparent gain in mass it is because something (usually a gas) has been gained.
- During a chemical reaction there is a rearrangement of the atoms but none are lost or gained.

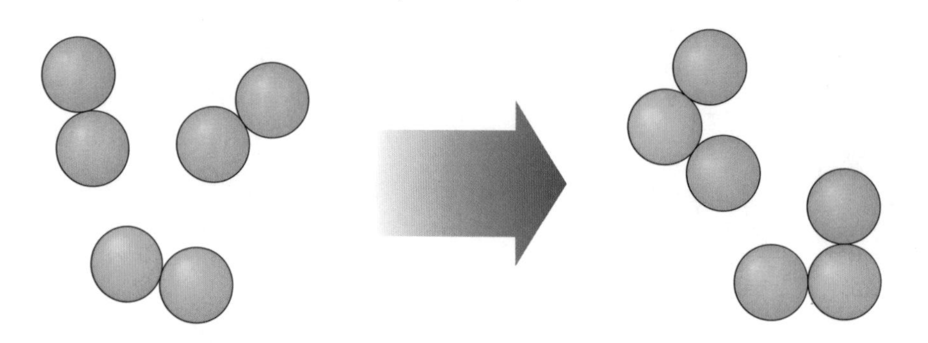

Fig. 6.12 The rearrangement of atoms that takes place when hydrogen and oxygen react to form water.

Progress Check

1 Write word and symbol equations for the reaction of magnesium and oxygen.

2 Write a symbol equation for the reaction of hydrogen and oxygen to produce water.

3 When a marble chip is added to dilute hydrochloric acid a reaction takes place.

$$CaCO_3(s) + 2HCl(aq) \rightarrow CaCl_2(aq) + H_2O(l) + CO_2(g)$$

There seems to be a loss of mass when the reaction takes place in a test tube. Explain why this is so and how you could show that, in fact, there is no mass change.

1. Magnesium + oxygen → magnesium oxide $2Mg(s) + O_2(g) \rightarrow 2MgO(s)$ 2. $2H_2(g) + O_2(g) \rightarrow 2H_2O(l)$

3. The apparent loss of mass is due to carbon dioxide escaping. If apparatus could be devised so carbon dioxide was not allowed to escape, the mass would be unchanged.

Practice test questions

The following questions test levels 3-6

(a) This question is about making zinc sulphate crystals.

Read the passage and answer the questions that follow.

Zinc carbonate powder was added to a colourless liquid **(A)** until no more bubbles of gas were produced when fresh zinc carbonate was added.

The excess zinc carbonate was removed from the solution.
The solution was concentrated by heating and then left to cool. Crystals of zinc sulphate were formed.

(i) Write down the name of the gas produced. **[1]**

....

(ii) Write down the name of the liquid **A**. **[1]**

....

(iii) Why were no more bubbles of gas produced? **[1]**

....

(iv) What method was used to remove the excess zinc carbonate from the solution? **[1]**

....

(b) The table shows results obtained when zinc oxide and zinc carbonate are heated in separate test tubes.

	Appearance before heating	**Appearance after heating**	**Mass before heating in g**	**Mass after heating in g**
Zinc oxide	White powder	Turns yellow but turns white again on cooling	2.3 g	2.3 g
Zinc carbonate	White powder	Turns yellow but turns white again on cooling	2.3 g	2.0 g

(i) Give reasons why the information suggests zinc carbonate is split up but zinc oxide is not. **[2]**

....

....

(ii) Finish the word equation. **[2]**

Zinc carbonate → .... + ...

The following questions test levels 4-7

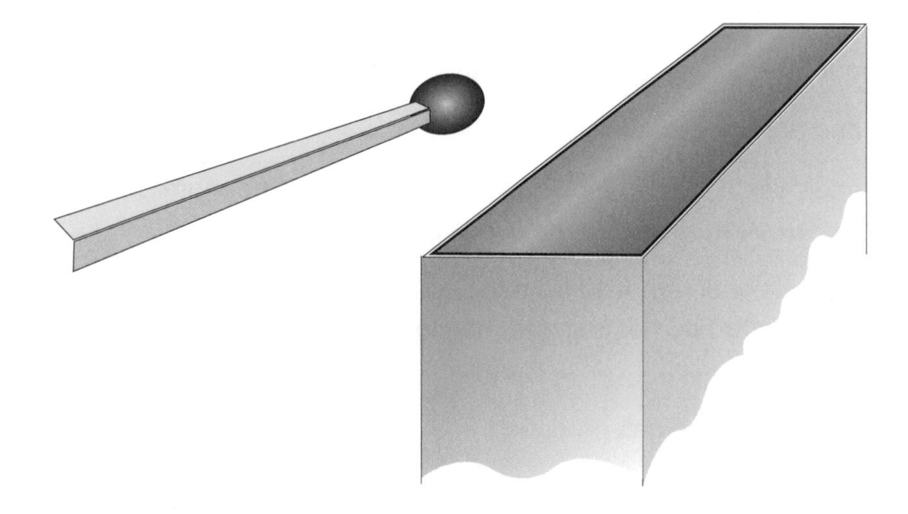

The diagram shows a match and the side of a matchbox.
The match head consists of potassium chlorate, $KClO_3$, sulphur and carbon.

(a) What is the job of the potassium chlorate? **[1]**

....

(b) Why is it necessary to strike the match on the side of the box? **[1]**

....

(c) Write symbol equations for the burning of sulphur and the burning of carbon. **[2]**

....

....

(d) What is produced when wood burns in a plentiful supply of air? **[2]**

....

Physics

Chapter Seven (Year 7)		Studied	Revised	Practice Questions
7.1 Energy resources	Fuels and energy Renewable resources Measuring energy			
7.2 Electric circuits	Current in circuits Different types of circuit Controlling the current Measuring current Changing the current			
7.3 Forces and their effects	What do forces do? Staying still Floating and sinking Getting moving Resistive forces			
7.4 The Earth and beyond	The view from Earth The view from outside			
Chapter Eight (Year 8)				
8.1 Heating and cooling	Heat and temperature Moving energy around Insulation			
8.2 Magnets and electromagnets	Magnetism Magnetic fields Electromagnetism Using electromagnetism			
8.3 Light	How light travels Seeing Mirrors More images Colour Colour addition Colour subtraction			
8.4 Sound and hearing	Sound production Travelling sound Hearing			
Chapter Nine (Year 9)				
9.1 Energy and electricity	Energy transfer Energy flow Energy in circuits Measuring voltage Generating electricity			
9.2 Gravitation and space	Gravitational force Planetary orbits Satellites			
9.3 Speeding up	Who won the race? How long does it take? Using graphs			
9.4 Pressure and moments	Under pressure Quantifying pressure Turning forces A question of balance			

The topics covered in this chapter are:

- Energy resources
- Electric circuits
- Forces and their effects
- The Earth and beyond

7.1 Energy resources

After studying this topic you should be able to:

- understand how energy can be released from a fuel
- describe how renewable energy resources can be used for heating and to generate electricity
- explain why it is important to conserve fossil fuels
- identify and describe some energy transfers

Fuels and energy

Even sitting doing nothing needs energy to keep our hearts pumping blood round our bodies.

An energy resource is a supply of energy.

Whatever we do needs **energy**. Energy is needed to:

- make things **move**
- **heat** our living space
- **light** the areas where we live and work

Electricity provides much of the energy that we need each day. The diagrams show examples of electricity being used to produce heat, light and movement.

Electricity is used to produce movement in the motor of this vacuum cleaner.

Electricity is used to heat this grill.

Electricity causes this lamp to emit light.

Fig. 7.1

Some modern power stations generate electricity by burning wood from fast-growing willow trees.

It is important to remember that all the energy stored in coal, oil and natural gas originally came from the Sun.

Most of our electricity is generated by burning a **fuel**. A fuel is something that burns in air and releases heat. Common fuels include wood, gas, coal and oil. Of these, wood is easy to replace as more trees can be grown. Coal, oil and natural gas are **fossil fuels** – they took millions of years to form and cannot be replaced.

Coal was formed from giant fern-like plants. They trapped energy from the Sun as they grew. In time the decaying remains of these plants became covered in mud, sand and clay. Over millions of years the pressure from above, together with heat from the Earth below, caused coal to form from the remains of the ferns.

Oil and natural gas formed in a similar way, but from animals rather than plants. When animals that live in the sea die, their remains form layers on the seabed. The combined effects of pressure and heat over millions of years resulted in oil and gas.

Fossil fuels are being used up rapidly.

- The UK has enough coal to last for 200 years.
- A lot of this coal will be expensive to recover from the ground.
- The known reserves of gas and oil will not last as long as coal, but new reserves are being found every year.
- Our use of gas, oil and coal needs to be as efficient as possible so as not to waste these fuels.
- We need to find alternative sources of energy for transport and for generating electricity.

Progress Check

1 Which of these are fuels?

electricity gas light petrol

2 Complete the sentence:

A fuel is a material that ________ and produces ________.

3 Which fossil fuel was formed from the remains of plants?

1. Gas and petrol 2. burns; heat 3. Coal

Renewable resources

We will always need to heat and light our homes and to have energy for transport. Fossil fuels will not last for ever, so we have to develop the use of renewable energy resources.

A renewable energy resource is one that will not run out during the lifetime of the Earth.

A common error at Key Stage 3 is to describe a renewable resource as one that can be re-used over and over again. Wood can only be burned once!

Plants are renewable energy resources that provide food, which in turn provides the energy for our bodies. They are renewable because we can grow fresh crops each year. Energy from plants is known as biomass. Other renewable resources include:

- wind
- waves and tides
- moving water
- the Sun
- geothermal energy

Fig. 7.2 A wind farm.

Energy from the Sun makes the **wind** blow. The energy in moving air has been used for thousands of years to turn windmills. Recently in the UK there has been a large increase in the number of wind turbines that generate electricity. These can be built singly or in groups, called wind farms. In some areas of the country the wind blows all the time. There are now plans to build wind farms off the east coast, so that people are not affected by the noise that they make.

The Sun also has some effect on the tides. Very high tides are caused by the Sun and the Moon pulling together.

The sea provides two important energy resources: **waves** and **tides**. Waves are caused by the wind, and tides are due mainly to the effect of the Moon pulling on the oceans. Attempts to extract energy from the waves have proved costly and unreliable, as the equipment has been damaged in storms.

Energy from tides has been used for hundreds of years to drive mills. It is predictable and reliable, as it depends only on the Moon orbiting the Earth. In the UK there are no power stations driven by tides, but the one on the estuary of the river La Rance in France generates enough electricity to supply a large town.

In Scotland and Wales there are fast-flowing streams and rivers that are used to generate electricity from **moving water**. The water passes through a **turbine**, which drives a **generator**. The energy source here is the Sun, which causes the evaporation of water from the sea. This water vapour forms clouds and falls as rain.

Most of our energy comes from the Sun. The exceptions are energy from the tides, which comes mainly from the Moon, and geothermal energy.

Energy from the Sun is used directly in three different ways:

- It is used to grow crops.
- It can be used to heat water.
- It can be used to generate electricity.

In countries with more sunshine than in the UK, hot water tanks on the roofs of houses are a common sight. Energy from the Sun is used to heat water as it passes through copper pipes. These pipes are painted black to absorb as much radiant energy as possible.

Fig. 7.3 This telephone box uses solar cells to provide the electricity that it needs.

Solar cells use energy from the Sun to produce electricity. They are expensive to make so the cost of the electricity from them is high. Another disadvantage is that they only produce electricity in daylight. They are useful to power calculators, which only use a small amount of electricity. They can also be much cheaper to install than mains electricity in remote places. In some countries they provide the electricity for telephones. A battery is used to store electricity generated during the day so that the telephones can be used at night.

Solar cells are used by satellites and spacecraft to generate electricity. They have vast panels of cells to generate enough electricity for the on-board computers and other devices. They also store the energy in batteries so that they have a reserve supply.

Geothermal energy is the energy in hot rocks below the Earth's surface. The energy in the rocks is used to heat water. If the rocks are hot enough, they can be used to generate steam to drive a turbine and produce electricity.

Measuring energy

When a gas or electricity bill arrives at home, the cost is for the energy transferred through the gas pipes or electricity cables.

One **joule** of energy is a very small amount. It is the amount of energy that leaves your body when you lift a 1 kg bag of sugar through a height of 1 metre.

Key Point

Energy is measured in joules (J).

The cost of boiling a full kettle of water is about 1p!

To boil a kettle of water from cold needs about 600 000 J of energy, so you can imagine that a lot of energy is needed to heat enough water to fill a bath.

Progress Check

1. Wood is renewable because it can be re-used. True or false?
2. What name is given to energy obtained from hot underground rocks?
3. Most of the energy used to boil a kettle of water comes from fossil fuels. True or false?
4. If it takes 10 J of energy to lift a 1 kg weight to a height of 1 m, how much energy is needed to lift a 25 kg sack of potatoes a vertical distance of 3 m?

1. False 2. Geothermal energy. 3. True 4. 750 J

7.2 Electric circuits

After studying this topic you should be able to:

- **recall the symbols used in a circuit diagram**
- **state that the current is the same at all points in a series circuit**
- **understand how current divides along the branches of a parallel circuit**
- **explain how the current in a circuit depends on the voltage and the resistance**
- **compare the advantages of series and parallel circuits**

Current in circuits

For a current to pass, the positive and negative terminals have to be joined by a conductor.

Using a lamp and a dry cell or a low voltage power pack, you can easily see that electricity is not like water or gas that flow along a pipe and come out when a tap is opened.

You need to have a complete route from the cell to the lamp and back again. This route is called a **circuit**.

Key Point

In a circuit there is a complete current path from the positive terminal of the battery or power supply to the negative terminal.

Remember, a conductor allows electricity to pass through it but an insulator does not.

If there is a break in the circuit, the lamp goes out. You can use a circuit with a break in it to test which materials are **conductors** and which are **insulators**.

Air is normally an insulator, so an air gap in a circuit can be used as a switch.

- If the gap is closed using a conducting material, the circuit is switched on.
- The circuit is switched off when the gap is open.

A circuit diagram is a shorthand way of showing how to connect the components in a circuit.

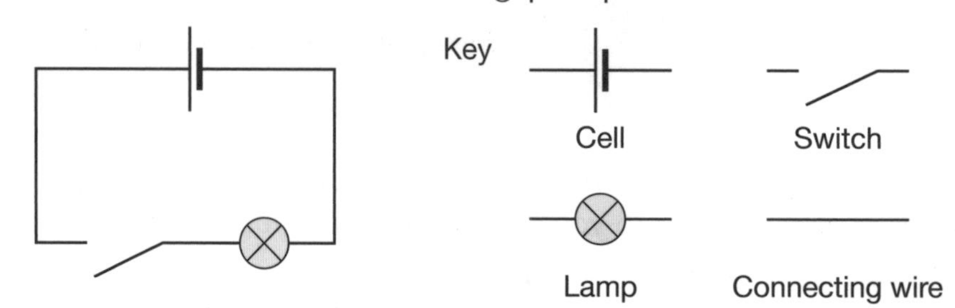

Fig. 7.4 This diagram uses circuit symbols to show a lamp that can be switched on and off.

Here are some more symbols that you may find useful.

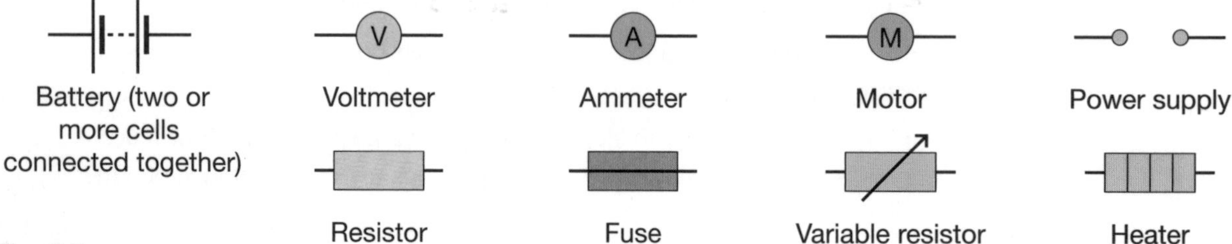

Fig. 7.5

Progress Check

1 Complete the sentence:

Electricity is used to _____ things up, make things ________ and make things give out ________.

2 An insulator allows electricity to pass through it easily. True or false?

3 In a circuit there is a complete conducting path from the positive to the negative terminal of the battery or power supply. True or false?

1. heat; move; light 2. False 3. True

Different types of circuit

There are two different ways of connecting two lamps to one cell or power supply:

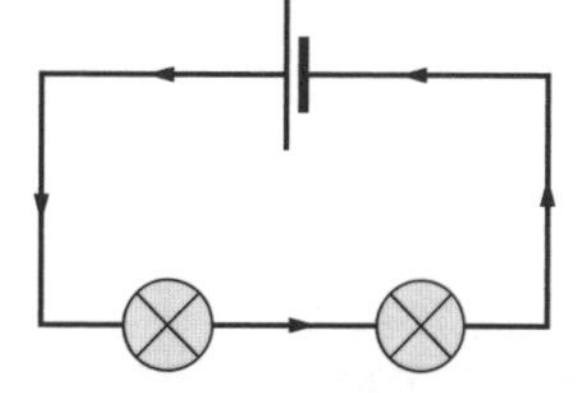

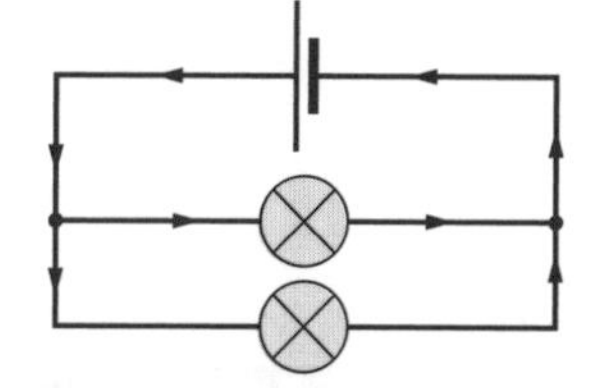

Fig. 7.6 A series circuit and a parallel circuit.

- The circuit on the left is called a **series** circuit. In a series circuit the current from the cell passes through each lamp in turn, one after the other.
- A **parallel** circuit is shown on the right. In a parallel circuit the current splits at the junction before the lamps and rejoins at the junction after the lamps.

Electronic systems, such as those in computers, radios and televisions, contain many series circuits but most household mains appliances are in a parallel circuit. The exception is some types of Christmas tree lights.

The components in a series circuit are either all turned on or all turned off.

In a series circuit, a break anywhere in a circuit turns the whole circuit off. With a parallel circuit, switches can be put in the branches of the circuit so that each switch controls just one device.

Controlling the current

Make sure that you know the circuit symbols for common components, such as lamps and variable resistors.

The brightness of a lamp shows the size of the electric current passing in a circuit; the brighter the lamp, the greater the current passing. The current can be made bigger or smaller if a **variable resistor** is included in the circuit.

The diagram shows a common type of variable resistor and the circuit diagram shows how it is connected in a lamp-dimming circuit.

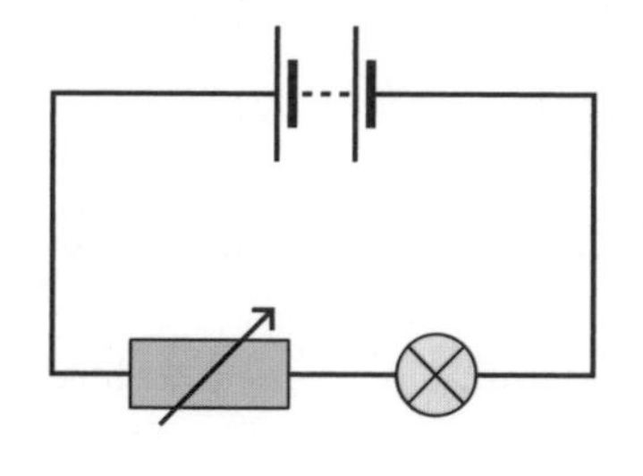

Fig. 7.7 A variable resistor connected in a lamp-dimming circuit.

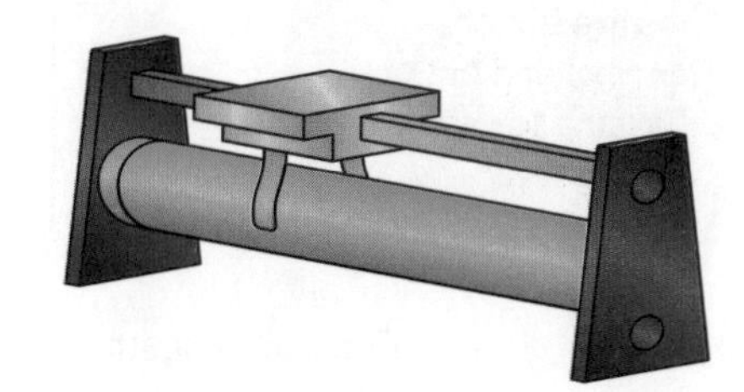

Moving the slider to one end of the variable resistor makes the lamp light at its brightest. The lamp is dimmest when the slider is moved to the opposite end.

Variable resistors can control other things besides lamps. They act as the volume control on radios and they can also be used to control the speed of an electric motor.

Progress Check

1 (a) Which type of circuit would you use for lighting your bedroom and that of your sister, series or parallel?

(b) Explain why this circuit should be used.

2 If you increase the resistance of a variable resistor, what happens to the current passing in it?

3 Complete the sentence:

Decreasing the resistance in a circuit causes the current to ___________.

1. (a) Parallel (b) So that each lamp can be switched on and off independently of the other lamp.
2. It decreases. 3. increase

Measuring current

A lamp 'blows' when part of the filament becomes so hot that it melts, breaking the circuit.

Sometimes the current passing in a circuit may be so small that it is not enough to light a lamp. Or it may be too large and would cause the lamp to 'blow'.

An **ammeter**:

- detects a greater range of current than a lamp does
- gives more precise measurements and enables you to make comparisons

Key Point

An ammeter is a device that measures the size of an electric current in amps (A). Ammeters are always connected in series in a circuit.

Ammeters are either digital, which are easy to read, or analogue, which involve a needle moving over a scale. Analogue meters call for more care when taking readings, as you have to interpret the scale divisions and make judgements about readings between scale divisions.

The diagrams below show how to connect an ammeter to measure the current passing into and out of a lamp.

The direction of a current is always taken as being from positive to negative. This means that current passes into the lamp from the positive battery terminal and out of the lamp towards the negative terminal.

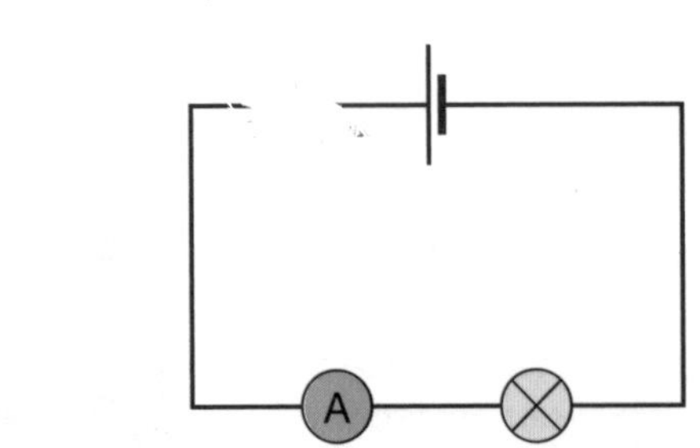

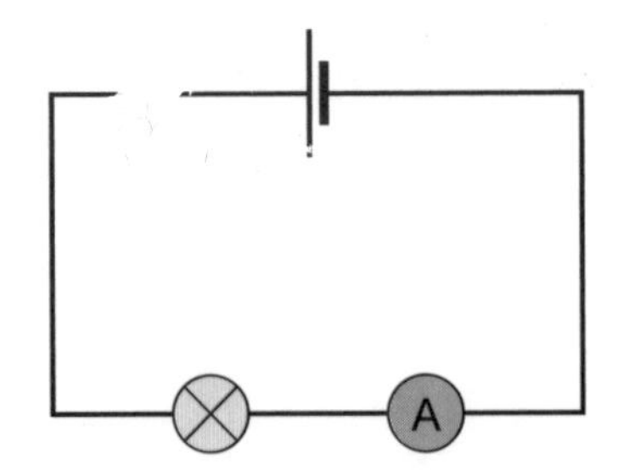

Fig. 7.8 Measuring the current going into a lamp... and the current going out.

Most people are surprised when they do this experiment and find that **the current coming out of the lamp is the same as that going in**. The lamp does not use up any current at all. The results obtained from the next circuit confirm this.

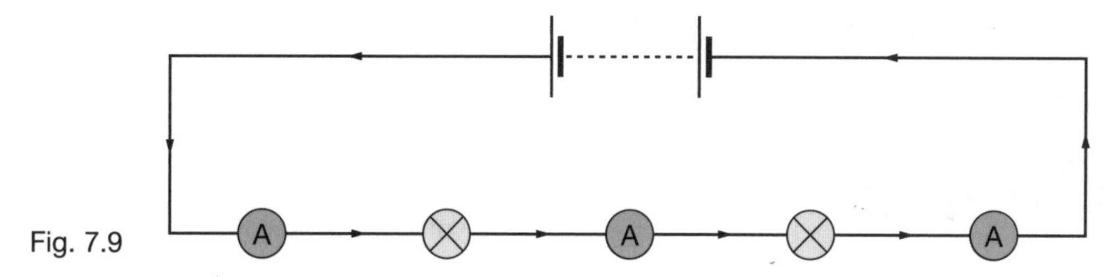

Fig. 7.9

All the ammeter readings are the same; the lamps have not used any current.

Key Point

The current is the same at all points in a series circuit.

This may seem strange because without a current the lamps do not work. So how do lamps produce light?

- What comes out of a lamp is **energy**, in the form of heat and light.
- The current in a circuit transfers energy from the source, the battery or power supply, to the lamp and other components, such as motors or heaters.

Energy is transferred around a circuit by moving charged particles. In metals the moving particles are electrons; these carry a negative charge. Electrons cannot move in an insulator but they are free to move around in a conductor. The electron movement is from negative to positive, even though we always mark current directions as being from positive to negative.

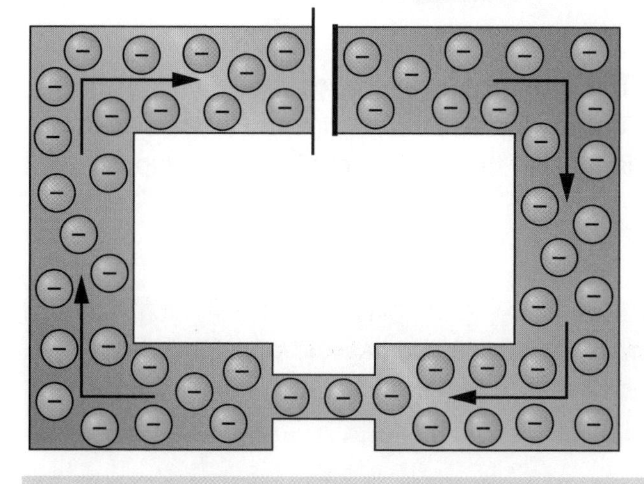

The diagram shows a model of charge flow in a circuit containing a cell and a lamp. The narrow part of the circuit represents the lamp filament.

Fig. 7.10 Charge flow in a circuit containing a cell and a lamp.

Progress Check

1. What instrument is used to measure current?
2. What is the correct unit of electric current?
3. Complete the sentence:

   The job of an electric current is to transfer ______________.
4. What is the sign of the charge on an electron?

1. An ammeter. 2. Amps 3. energy 4. Negative

Changing the current

In the section 'controlling current' you learned how a variable resistor can be used to vary the current in a circuit. The size of the current that passes in a circuit depends on:

> Resistance describes the opposition to electric current.

- the **voltage** of the current source
- the **resistance** of the circuit

A variable resistor works by changing the resistance. The more resistance there is, the smaller the current that passes.

When you increase the voltage that drives the current in a circuit, a bigger force acts on the moving charges. The charge travels round the circuit at a greater rate, so increasing the current.

The diagram shows that when you add more lamps in a series circuit, the current becomes less because you are increasing the circuit resistance.

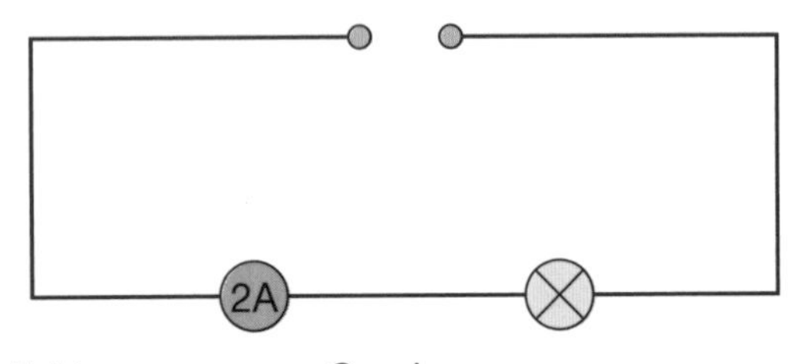

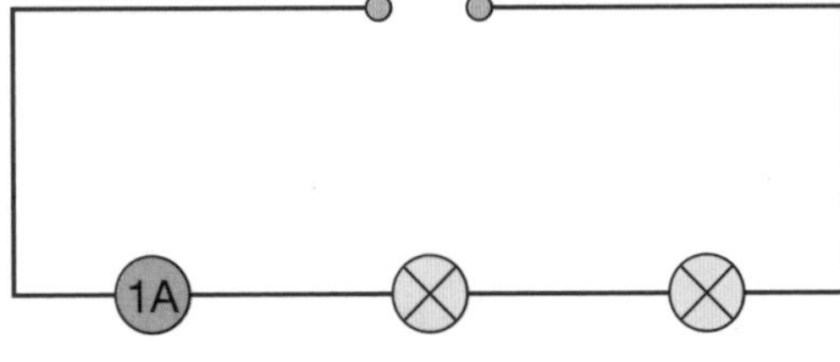

Fig. 7.11 One lamp Two lamps in series

> When lamps are added in parallel the circuit has less resistance as more current can pass.

Adding more lamps in parallel causes more current in the circuit because there are more routes available for the current to pass through, as the next diagram shows.

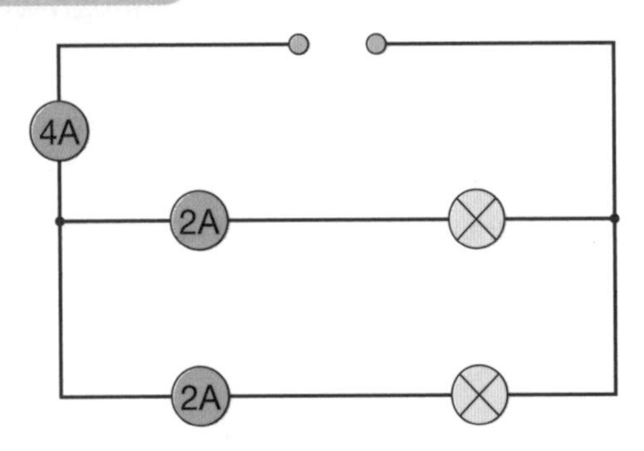

This circuit shows that when the current splits at a junction the sum of the currents along the branches is equal to the current from the power supply.

Fig. 7.12 Two lamps in parallel.

Key Point

In a parallel circuit, the current that passes into a junction is equal to the current that passes out.

Progress Check

1. What happens to the current in a circuit when more lamps are added in series with the power supply?
2. What happens to the current in a circuit when more lamps are added in parallel with the power supply?

1. It decreases. 2. It increases.

7.3 Forces and their effects

After studying this topic you should be able to:

- **draw arrows on diagrams to show forces**
- **describe forces as being due to one object acting on another**
- **understand that weight is a force that acts on a mass**
- **decide whether forces are balanced or unbalanced**
- **explain how resistive forces act against motion**

What do forces do?

Forces are acting everywhere. No matter where you look, you will see evidence of things **pushing** and **pulling** other things.

Forces can start and stop things from moving, cause changes in direction and change the shape of things when they squash or bend or stretch or twist them.

You should always use a phrase like this when describing a force.

Key Point

Forces are described by a phrase such as object A pulls or pushes object B.

When you draw forces on a diagram, you should use an arrow to show the direction of the push or pull. Here are some examples:

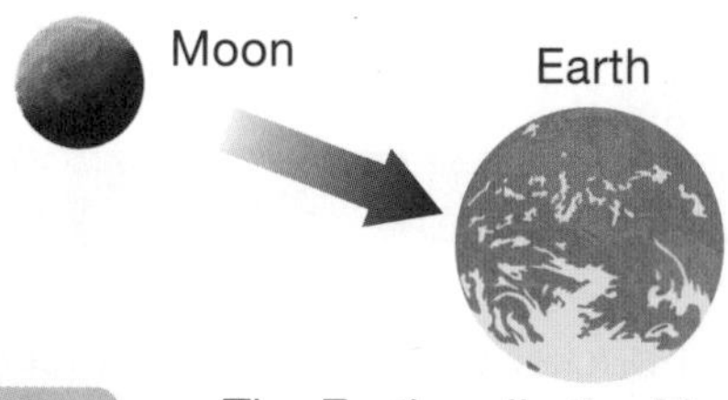

The Earth pulls the Moon.

The man pushes the pram.

The woman pulls the door.

Fig. 7.13

A common error is to describe the Earth's pull as 'gravity'. This is wrong; it is a gravitational force – the type of force that acts between masses.

- The Earth, like all other very massive objects, pulls everything else towards it.
- The force that pulls an object down towards the Earth is called its **weight**.
- Like all other forces, weight is measured in **newtons** (N for short).

Key Point

Weight is the downward pull of the Earth on an object.

On Earth the weight of each kilogram of material is about 10 N, so a 25 kg sack of potatoes is pulled towards the Earth with a force of 250 N.

The Moon also pulls things towards it but with a smaller force; on the Moon the potatoes would weigh about 38 N.

Staying still

Look around you – you are surrounded by things that are not moving. Perhaps you are sat at a desk that is not moving, or you may be sitting in a comfortable chair that is not moving. Even if you are reading this sat on a moving bus, you can look out of the window and see things that are not moving.

Everything that you can see has at least one force acting on it – the Earth's pull.

A common question in KS3 tests is to compare two forces acting on an object that is not moving. They are balanced – equal in size and opposite in direction.

If something is not moving there must also be another force pushing or pulling it so that the forces are **balanced**.

When you sit on a chair it changes shape; the springs and cushion get squashed so that they push up on you. You now have two equal-sized forces acting on you in opposite directions – the forces on you are balanced.

Here are some more examples of balanced forces.

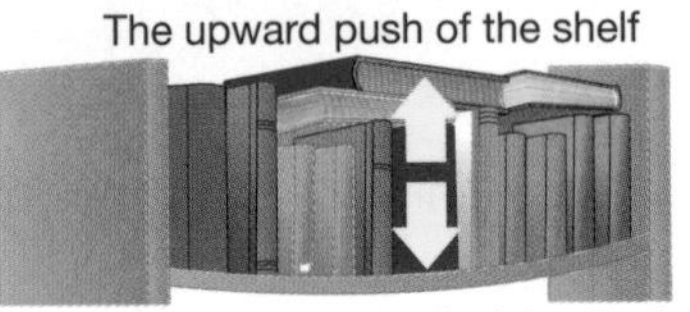

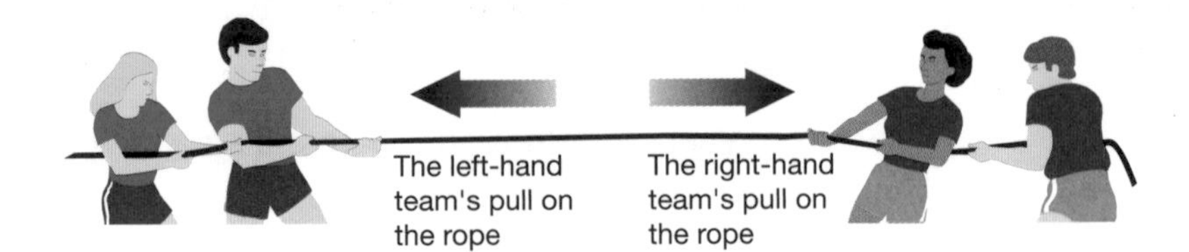

Floating and sinking

The correct description of this force is 'the water pushes the ball'.

- If you push a plastic ball into a bucket of water, you can feel the water pushing it back up – the more you push it in, the bigger the upward push of the water becomes.
- When a ball floats the forces on it are balanced; the upward push of the water is equal in size to the downward pull of the Earth.
- An object sinks if its weight is greater than the upward push of the water.

A heavier ball has to displace more water because it needs a bigger upward force to balance its weight.

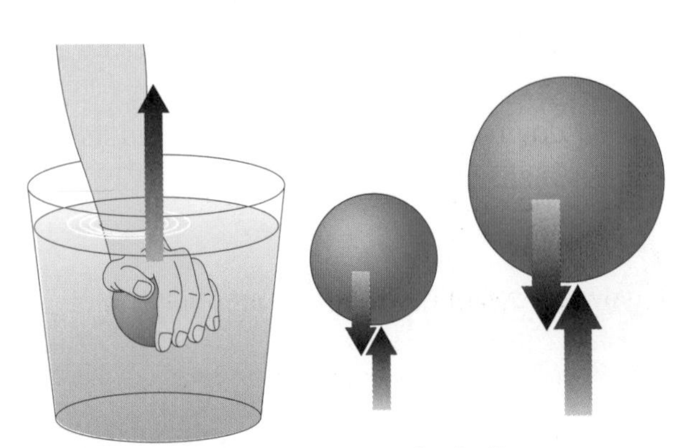

Fig. 7.15 The water pushes up on the ball.

Air also pushes up on things; it is the upward push of the air that causes a hydrogen-filled balloon to rise.

A hydrogen-filled balloon rises because the upward push of the air is greater than the downward pull of the Earth.

Progress Check

1. If two forces acting on an object are equal in size and opposite in direction, are they balanced or unbalanced?
2. Complete the sentence:

   When you sit on a chair, the downward pull of the ________ is balanced by the __________ push of the __________.
3. Describe the two forces that act on a ball floating on water. How can you tell that these forces are balanced?

1. Balanced 2. Earth; upward; chair.
3. The Earth pulls the ball down. The water pushes the ball up. The ball is not moving.

Getting moving

Key Point

To start or stop movement, to speed up or slow down, there needs to be an unbalanced force.

Fig. 7.16

A common error is to describe air resistance as friction. This is wrong. Frictional forces oppose objects from sliding over each other – see page 105.

A train, a car, a bus, a bike and a person walking all need a force to make them move. Can you identify the forces shown in the diagrams of the cyclist? The one on the left is the **driving** force that pushes the cycle along; the one on the right is the **air resistance** that acts against the cyclist's motion. Anyone who has ever ridden a bicycle has felt the effect of air resistance; the faster you go, the bigger this resistive force gets as you have to push more air out of the way each second.

When you are riding a bicycle there are always resistive forces.

- Most of the resistance to motion comes from the air.
- There are also resistive forces in places such as the wheel bearings that oppose a cyclist's motion.

Because of these resistive forces, you have to keep pedalling just to maintain a steady speed. Putting the brakes on creates an extra **resistive force** so that you slow down more rapidly.

These are all examples of unbalanced forces.

The driving force and the resistive force are both at work when a cyclist is pedalling.

- For the cyclist to speed up, the driving force needs to be bigger than the resistive force.
- If the resistive force is bigger than the driving force, the cyclist slows down.
- Cyclists usually stop pedalling when they brake, so there is no driving force, but only the resistive force acting.

When you get on a bicycle and start to pedal the resistive force is small at first so you speed up quite rapidly.

- As your speed increases so does the resistive force.
- Eventually you get to a speed where the resistive force is equal in size to the driving force.
- When this happens you stop speeding up and travel at a constant speed.

The forces on an object are balanced when it is not moving or not changing its speed or direction.

The forces acting on an object moving in a straight line at a constant speed are balanced. They are the same size but act in opposite directions so that their combined effect is just as if there was no force acting at all.

Fig. 7.17

Unbalanced forces are needed to cause any change in motion. A change in direction needs an unbalanced force. The diagram shows a girl on a trampoline. To make her move up when she bounces, the upward push of the trampoline has to be bigger than the downward pull of the Earth.

When the girl has lost contact with the trampoline and is moving upwards, the Earth's pull is unbalanced. This causes her to slow down and change direction, making her fall again.

The diagram shows the horizontal forces on a car that is travelling at a steady speed.

1 What does the arrow labelled A represent?

2 What does the arrow labelled B represent?

3 How can you tell that the car is travelling at a steady speed?

4 Draw a diagram to show the forces on a car that is speeding up.

5 Draw a diagram to show the forces on a car that is braking.

1. The resistive forces on the car. 2. The driving force. 3. The arrows are the same size, showing equal-sized forces. 4. The diagram should show the driving force arrow longer than the resistive force arrow. 5. The diagram should only show a resistive force arrow.

Resistive forces

When you swim you have to work to push against the water. When you run you have to work to push against the air. You can probably run faster than you can swim because the resistive force from the air is less than that from the water.

There are other resistive forces that act on moving objects. Sometimes their effect is to slow down or stop the motion but they also make motion possible. Without resistive forces we cannot walk, and bikes, cars, buses and trains cannot move.

Key Point

Friction is a resistive force that acts against sliding or slipping.

- If you push a book across a desk, **friction** is the force that slows it down and stops it.
- Friction always acts in the opposite direction to any sliding motion.
- Rough surfaces cause bigger friction forces than smooth ones do, so if you want something to slide you should keep the surface smooth and polished or lubricated.

Ice is a good surface to slide on but a very poor surface to walk or ride a bike on.

It may seem strange that we have to push backwards to move forwards. You also do this when you push away from the side of a swimming pool.

When we walk we rely on friction to stop our feet from slipping:

- To walk forwards our feet push backwards on the ground.
- The friction force stops them from moving backwards and pushes us forward.
- Without friction, our feet would just slip and we would not get anywhere!

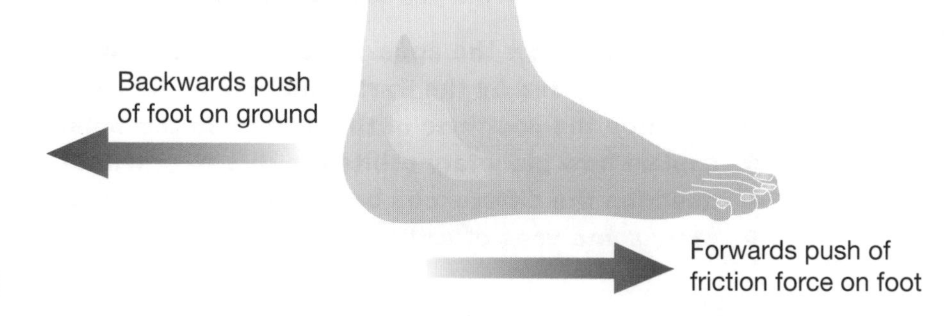

Fig. 7.18

Trains also need friction between the wheels and the rails. When wet leaves get on the track the friction force is reduced and the train wheels slip.

A common error is to state that the skydiver starts to move upwards when the parachute is opened.

Parachutists depend on resistive forces to slow them down. A skydiver who jumps from an aircraft speeds up to a speed of about 60 m/s. At this speed the air resistance balances the Earth's pull on the skydiver. Opening the parachute causes the air resistance to get bigger, so the forces acting on the skydiver are no longer balanced. The skydiver now slows down.

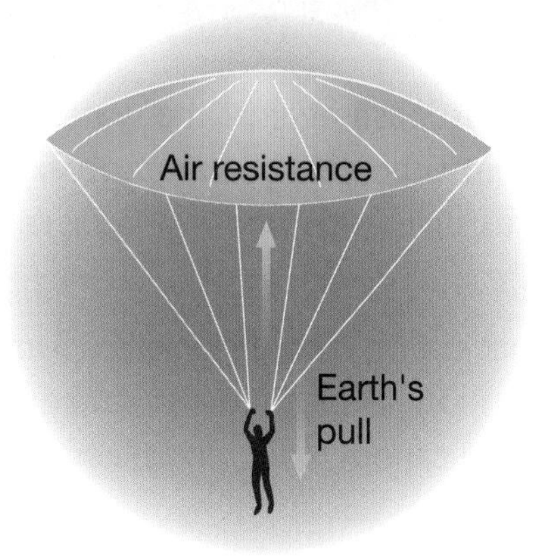

Fig. 7.19 Forces on a skydiver.

Eventually the skydiver falls at a much lower speed with the Earth's pull and the air resistance once again in balance.

Progress Check

1 When a parachutist opens her parachute, what happens to:
   (a) the upward force acting on her?
   (b) the downward force acting on her?
   (c) her speed?
   (d) her direction of travel?

2 When we walk, the force that pushes us forwards is due to friction between our feet and the ground. True or false?

3 'Friction stops things from moving.' Explain whether this is true or false.

1. (a) It increases. (b) It stays the same. (c) It decreases. (d) It stays the same. 2. True 3. False. Without friction people would not be able to walk and wheel-based vehicles would not be able to move.

7.4 The Earth and beyond

After studying this topic you should be able to:

- **understand how the apparent daily movement of the Sun and other stars is caused by the Earth spinning on its axis**
- **describe the positions of the planets in the Solar System**
- **explain how planetary orbits are due to gravitational attractive forces**
- **describe the differences between how the Sun and planets are seen**
- **state some uses of artificial satellites**

The view from Earth

If you are sat at home reading this, you do not have a sensation of movement. It is not obvious that our **planet** Earth is moving through space with a speed of thousands of metres per second. So it is not surprising that early astronomers made the mistake of thinking that the **Sun** moves around the **Earth**.

On a sunny day you can plot the Sun's path across the sky by watching the movement of a shadow.

- The shadow is longest in the morning and evening when the Sun is low in the sky.
- As the shadow moves round in the morning it gets shorter. It is at its shortest at noon when the Sun has reached its highest point in the sky.
- After noon, the shadow starts to lengthen again. You can see very long shadows near sunset on a sunny evening.

A common error is to confuse the Earth's rotation on its axis with its movement around the Sun.

These changes in a shadow seem to be caused by the Sun moving across the sky – rising in the East and setting in the West. The diagram shows how the Sun appears to move each day.

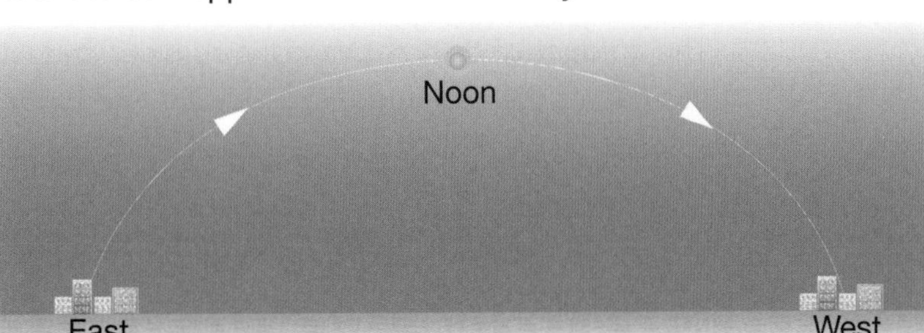

Fig. 7.20

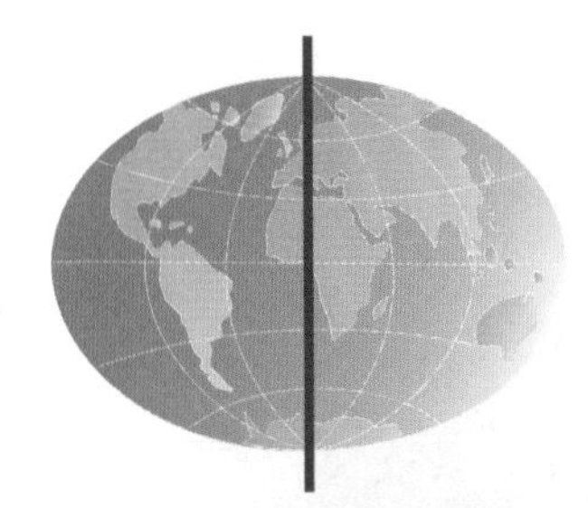

Fig. 7.21

We now know that this 'movement' of the Sun is actually due to the Earth turning round once each day. The Earth **spins** on its **axis**, an imaginary line going through the centre of the Earth from pole to pole.

The Earth makes one complete rotation on its axis in one day.

If you were out in **space**, looking down at the Earth's North Pole, you would see the Earth turning round in an anticlockwise direction.

Key Point

This daily rotation of the Earth causes day and night. It also causes the Sun's apparent movement across the sky.

The diagram shows Britain at sunrise, noon and sunset. This is the view you would have if you were looking down at the North Pole. The outer circle of the Earth is the equator.

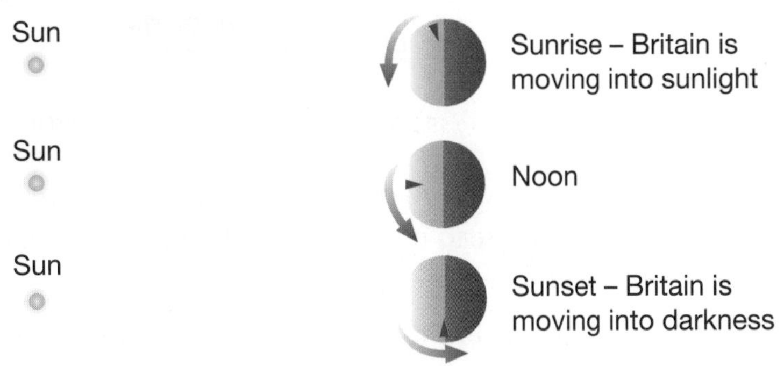

Fig. 7.22

Progress Check

1. Use the diagram to state where you would have to look to see the Sun at different times of day.
2. The Sun makes one revolution of the Earth each day. True or false?
3. Complete the sentence:

   The Earth takes ____________________ to spin once on its axis.

1. To the east at sunrise, overhead at noon and to the west at sunset.
2. False 3. One day or 24 hours.

The amount of daylight varies in **summer** and **winter**, and the highest position of the Sun in the sky also changes with the season.

- Twice a year, at the spring and autumn equinox, day and night are of equal length.
- Summer brings longer days than nights, with the Sun being higher in the sky.
- The diagram shows the Sun's apparent path at three different times of year.

Questions about the differences in the Sun's daily motion in winter and summer are common in tests and examinations.

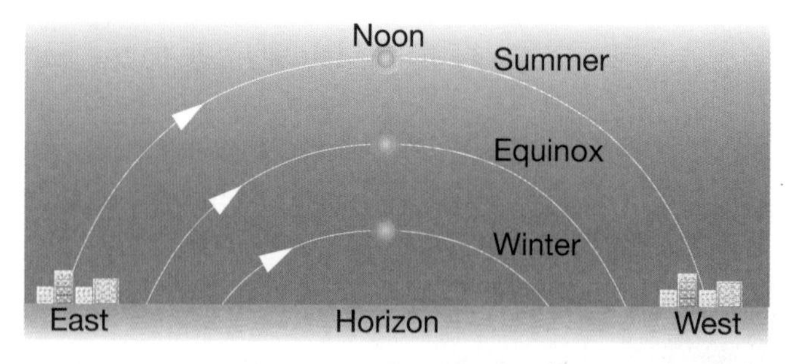

Fig. 7.23

The number of hours of **daylight** and **darkness** varies throughout the year. This is due to the Earth being **tilted**. The Earth's axis always points in the same direction, towards the star **Polaris**.

The Pole itself is light for 24 hours a day in the height of summer.

When it is summer in Britain the **northern hemisphere** is tilted towards the Sun, so we spend more than 12 hours in daylight. The nearer you are to the North Pole in summer, the longer your days are.

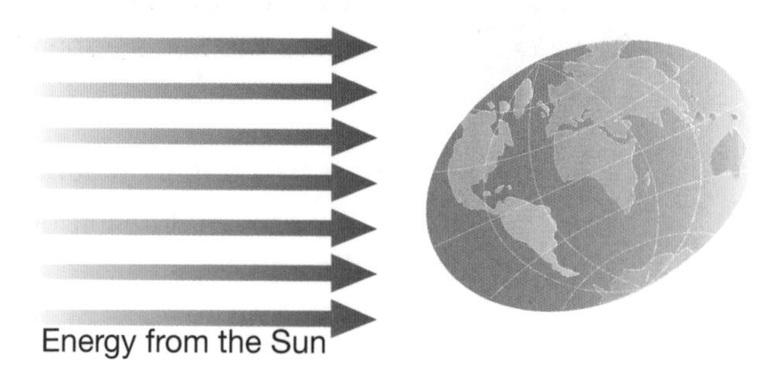

Fig. 7.24 Summer in the northern hemisphere.

This not only makes the days longer, it also means that energy from the Sun is spread over a smaller area than in winter, giving us a warmer climate in summer.

When it is winter in Britain, energy from the Sun is spread over a larger area.

Our Sun is the only **star** that we see during the day; the other stars are there but the bright light from the Sun prevents us from seeing them. Polaris, or the Pole Star, is a very bright star. As mentioned earlier, the Earth's axis points towards Polaris, so you can see it by looking in the sky directly north. The diagram shows the Pole Star as it can appear in the winter sky.

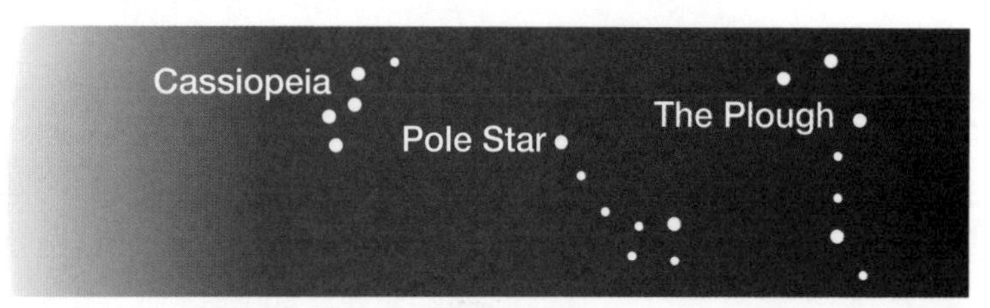

Fig. 7.25

The Pole Star is the one star in the sky that you can always see in the same place.

To see the effect of the Earth's rotation:

- stand under the ceiling light in the centre of a room
- imagine that this is the Pole Star; the Plough is to your right
- rotate yourself anticlockwise through a quarter of a turn – the Plough now appears to be 'above' the Pole Star

The Earth's spin on its axis makes the stars in the northern sky appear to revolve anticlockwise around the Pole Star. Stars in the southern sky appear to move round clockwise.

If the Earth's only movement was spinning on its axis, we would see the stars in the same place at the same time each night. But we have to take into account the movement around the Sun.

- As we revolve once around the Sun it looks as if the stars are turning once around the Pole Star.
- Since there are 365 days in a year and 360° in a circle, the pattern of the stars seems to move by about 1° each day.
- If you look at the stars at the same time of night, in one month they should have moved round in the sky by about 30°.

A constellation is a group of stars that makes a pattern.

In winter the constellation Orion can be seen in the southern sky. The left-hand diagram shows what it looks like one winter's night at 9 pm. In the right-hand diagram it has moved round through 30°, that is one twelfth of a rotation. The right-hand diagram could have been drawn at 11 pm on the same night or at 9 pm one month later.

Fig. 7.26

Progress Check

1 The diagram shows the Sun at midday in winter.

(a) Draw a line on the diagram to show the path that the Sun follows between sunrise and sunset.

(b) Write an **S** at the place where you would expect to see the Sun at midday in summer.

(c) What causes the apparent movement of the Sun across the sky each day?

2 What do the stars in the northern hemisphere appear to revolve around?

1. (a) The line should be a curve from the horizon in the east (left) to that in the west.
(b) The S should be above the position of the Sun shown in the diagram.
(c) The Earth's rotation on its axis. 2. The Pole Star (Polaris).

The view from outside

If you could look at the **Solar System** from a viewpoint where you could see all the planets, what would you see? The most striking object would be the Sun, shining with a brilliant white light. You would also be struck by the colours of the planets and the contrast between the rich orange colour of Venus and the blue planet Earth. Of course, you would only be able to see half of each planet, because planets are only visible by the sunlight that they **reflect** and only the half of each planet facing the Sun would be lit up.

There are a number of phrases that people use to remember the order of the planets. You could try making up your own.

You would notice four planets close to the Sun:

- Nearest to the Sun is the tiny planet **Mercury**, its surface covered in impact craters.
- Next comes **Venus**, its dense atmosphere reflecting only orange light.
- The third planet out is the planet **Earth**.
- The fourth planet out from the Sun, **Mars**, appears a blood-red colour.

From outer space you would not be able to see any evidence of human activity on Earth. You would, however, notice that the Earth has a **satellite** of its own – a moon very similar in size and appearance to Mercury. The Earth's moon takes about 28 days to complete an orbit of the Earth. It takes the same time to spin once on its axis, so the same side of the Moon always faces towards the Earth.

Mercury has the shortest orbit time of all the planets. It takes just 88 days to travel once around the Sun.

Looking at the planets, you would see that:

- they all go round the Sun in the same direction
- the speed of a planet depends on its distance from the Sun
- Mercury moves fastest of all and it also has the shortest distance to travel to complete an orbit

Moving your eyes further out, you would notice a lot of rocky fragments orbiting the Sun between Mars and the first of the outer planets, Jupiter. These fragments of rock, up to 100 km in diameter, form the **asteroid belt**.

The outer planets, in order from the Sun, are:

- **Jupiter** – the largest planet in the solar system. It has a spectacular appearance, with a swirling atmosphere and its Great Red Spot. It also has 16 moons in orbit around it.
- **Saturn**, the sixth planet, which is a very bright yellow object in the sky. It has more than 100 000 rings made up of dust, ice and rock. As well as its rings, Saturn has more than 20 satellites in orbit around it.
- **Uranus**, which also has rings and a total of 15 satellites, five of which are large moons.
- **Neptune**, the eighth planet, which is very similar in size and composition to Uranus. Being further away from the Sun, Neptune takes twice as long as Uranus to complete an orbit.
- **Pluto**, the outermost known planet, which still has to be photographed at close range. Pluto has one known moon, **Charon**. Pluto and Charon are very similar in size and mass, and they revolve around each other like a pair of ice skaters holding hands.

The scale of the Solar System is vast and it is very difficult to fit the Sun and planets on the same diagram. This diagram gives an idea of the relative sizes and distances of the planets from the Sun.

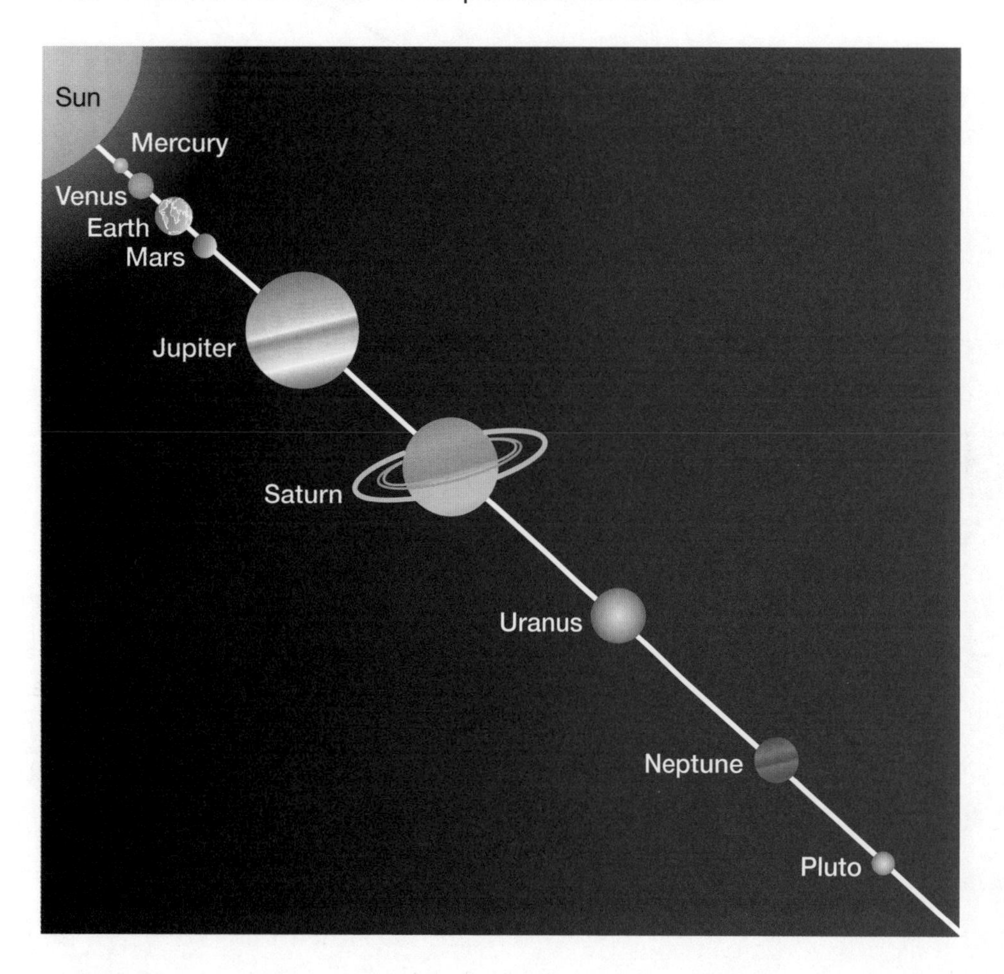

Fig. 7.27 Relative sizes and distances of the planets from the Sun.

Practice test questions

The following questions test levels 3-6

(a) Name two non-renewable energy sources. **[2]**

....

....

(b) Name two renewable energy sources. **[2]**

....

....

(c) The diagram shows a circuit that uses two switches to control a lamp.

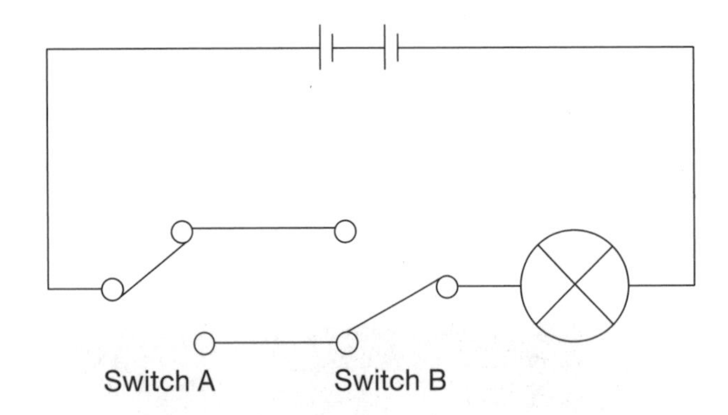

(i) Complete the table to show when the lamp is 'on' and 'off'. **[3]**

Switch A	**Switch B**	**Lamp**
up	down	off
up	up	
down	up	
down	down	

(ii) The switches are placed at the positions shown in the diagram.

Write down two ways of switching the lamp on. **[2]**

....

....

(iii) Suggest a use for this circuit. **[1]**

....

The following questions test levels 5-7

The diagram represents the Sun and the three planets closest to the Sun.

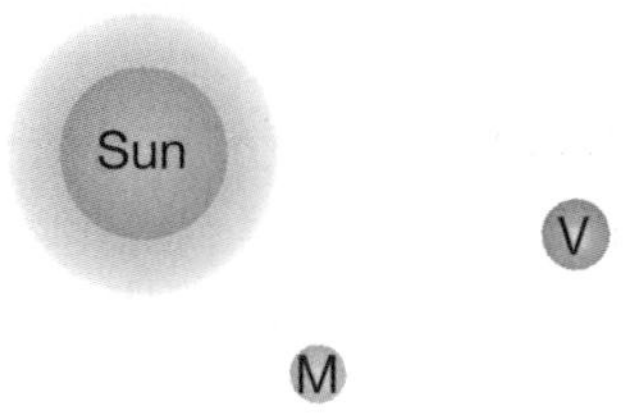

(a) Write down the names of the planets labelled M and V. **[2]**

....

(b) Which object shown on the diagram is a light source? **[1]**

....

(c) Venus can be seen from the Earth as a very bright object in the night sky.

Explain why Venus looks to be brighter than the other planets. **[2]**

....

....

(d) On the diagram, shade the part of the Earth that is in darkness. **[1]**

....

(e) Explain how the movement of the Earth causes day and night. **[1]**

....

Year 8

Chapter Eight

The topics covered in this chapter are:

- **Heating and cooling**
- **Magnets and electromagnets**
- **Light**
- **Sound and hearing**

8.1 Heating and cooling

After studying this topic you should be able to:

- **describe the factors that affect the temperature of an object**
- **explain how energy transfer takes place by the movement of particles and the emission and absorption of electromagnetic radiation**
- **explain how insulation is effective in keeping objects hot or cold**

Heat and temperature

Which is hotter: a bath full of hot water or the filament of a lamp that is giving out light? The answer is the lamp filament; its **temperature** is around 2000 °C, whereas that of the bath is about 50 °C. The diagram shows some typical temperatures measured on the **Celsius scale**.

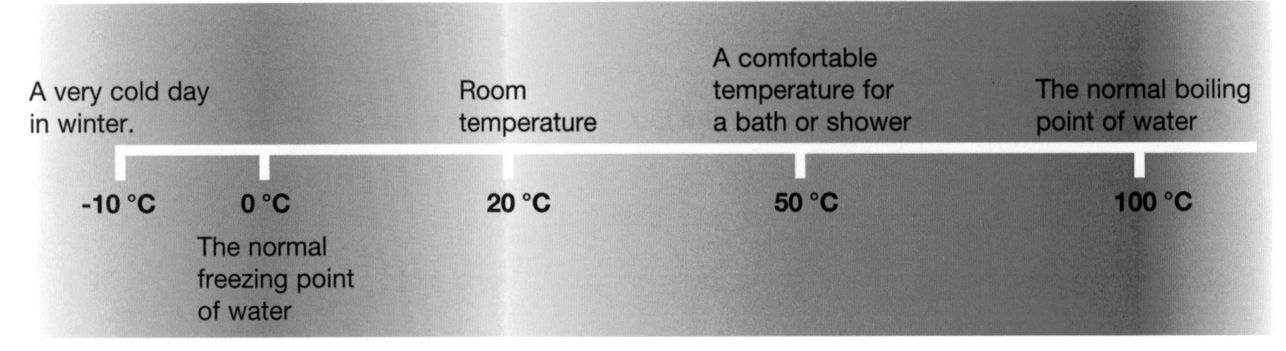

Fig. 8.1

Although the bath water is cooler than the lamp filament, it takes more **energy** from the gas or electricity mains to heat it up from cold. This is because:

- it takes more energy to heat 1 kg of water by each degree Celsius than it does to heat 1 kg of lamp filament by the same amount
- there is much more material to be heated – it takes several kg of water to fill a bath but only a fraction of a kg of tungsten to make a lamp filament

Temperature is a measure of hotness. It does not measure the amount of energy needed for an object to become hot.

Progress Check

1 Complete the sentence:

The water in a hot shower is ______________ than the glowing metal filament in a lamp but ______ energy is needed to heat it.

2 Suggest why a temperature of 0 °C is called the 'ice point'.

3 On bonfire night, the sparks from a sparkler are very hot. Suggest why they do not burn the skin when they land on a person.

1. cooler; more 2. Ice melts into water and water freezes into ice at 0 °C. 3. They lose very little energy as they cool down.

Moving energy around

If you leave a hot drink and an ice cube in the kitchen, the drink cools down but the ice cube warms up!

Key Point

Objects that are warmer than their surroundings lose energy; those that are cooler gain energy.

There are a number of ways in which energy moves from warm objects to cooler ones.

Energy is transferred through the walls of the mug by **conduction**.
In conduction:

- the particles in the warm, inner part of the wall have more energy than those in the cool outer part
- the particles in the warmer part vibrate with more energy than those in the cooler part
- energy is transferred from particle to particle through these vibrations

The diagram illustrates the process of conduction.

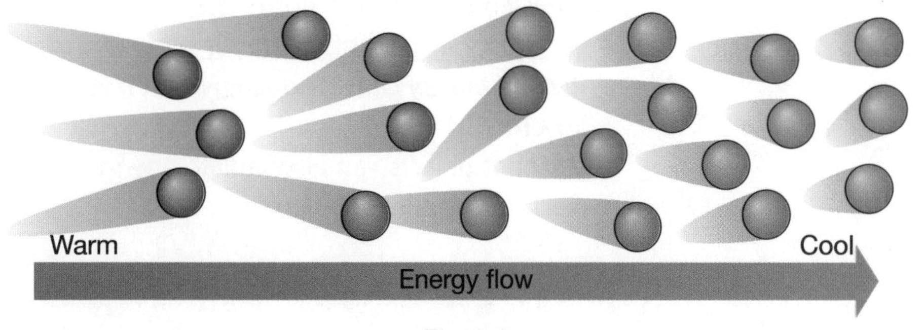

Fig. 8.2

Solids and liquids are better **thermal conductors** than gases. This is because their particles are closer together and can pass on the energy more readily.

Key Point

Metals are the best conductors because electrons can move through metals, taking energy with them to all parts of the metal.

When answering questions about convection currents, it is important to stress the changes in density.

Energy that flows through the walls is carried away by **convection currents** in the surrounding air.

- Air near the outside of the walls becomes heated.
- The warmed air expands and becomes less dense.
- The warm, less dense air rises and is replaced by cooler air from the surroundings.

Key Point

Convection currents are due to changes in density caused by heating and cooling.

A common error is to state that 'heat rises'. Heat is not a substance, so it cannot move. In convection currents, warm or cold liquid or gas rises or falls.

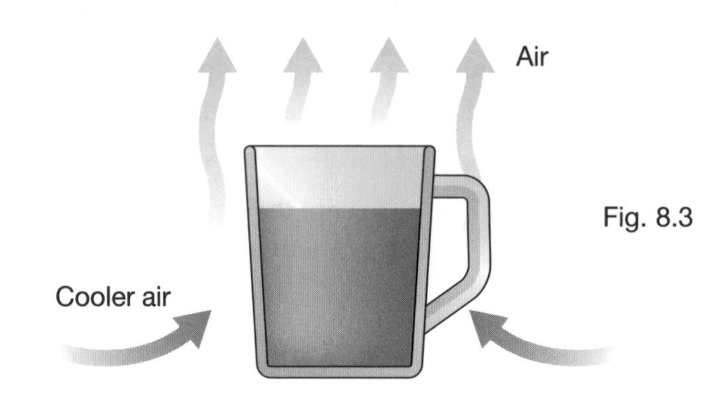

Fig. 8.3

As the diagram shows, convection also takes place at the surface of the liquid. Another way in which energy is lost from the surface is **evaporation**. When a liquid evaporates:

- the more energetic particles near the surface leave the liquid, becoming vapour
- this reduces the average energy of the particles remaining in the liquid
- the result of this is that the liquid loses energy and its temperature falls

All parts of the mug emit energy as **electromagnetic radiation**. Everything radiates energy, but the hotter the object the more energy it radiates.

- For most objects, this energy is **infra-red radiation**.
- **Radiant energy** causes cooling when it is emitted and heating when it is absorbed.
- Very hot objects can also radiate energy as light and other forms of electromagnetic radiation.

Progress Check

1. What type of electromagnetic radiation is emitted by **all** objects?
2. When a liquid evaporates, the least energetic particles leave the surface. True or false?
3. Which type of material is the poorest conductor of thermal energy: a solid, a liquid or a gas?

1. Infra-red 2. False 3. A gas.

Insulation

In cold weather, we put more clothes on to keep warm. In summer, we keep our picnic in a cool box. These are both examples of using **insulation**.

Key Point

The purpose of insulation is to reduce the energy transfer between an object and its surroundings.

Double-glazing uses the insulation properties of air trapped between two layers of glass to reduce heat loss from a house.

Most forms of insulation use air as the **insulator**. This is effective because air is very poor at conducting energy in the form of heat. Air is, however, very good at forming convection currents, so it needs to be trapped so that it cannot move.

Air is trapped between layers of clothing and pockets of air are trapped in foam insulation. This prevents energy transfer from taking place through convection currents and minimises energy transfer by conduction.

There are two other methods of minimising the energy transfer between hot and cold objects:

- If the object is liquid or contains a liquid, covering it reduces the energy loss through evaporation. This is very effective in keeping hot drinks hot and minimising the energy loss from a swimming pool when it is not being used.
- Wrapping an object in aluminium foil reduces the energy loss through infra-red radiation. Foil reflects radiant energy in the same way that a mirror reflects light. This is very effective in keeping food hot after it has been cooked.

Progress Check

1 Take-away food is sometimes wrapped in several layers of paper. How does this provide effective insulation?

2 Explain why aluminium foil is a good insulator for food that has been just taken out of an oven.

3 Covering a swimming pool with a polythene sheet at night reduces the amount of cooling. Explain why.

1. Air is trapped between each layer. Air is a good thermal insulator.
2. The aluminium foil prevents evaporation and reflects infra-red radiant energy back onto the food.
3. It stops evaporation from removing energy from the pool.

8.2 Magnets and electromagnets

After studying this topic you should be able to:

- **state which materials are magnetic and which are non-magnetic**
- **describe a magnetic field and interpret magnetic field patterns**
- **explain that there is a magnetic field associated with an electric current and how this is used in some devices**

Magnetism

A common error is to state that metals are attracted to magnets – this is only true for some metals.

Magnets attract iron, steel and nickel but they do not affect copper, brass, chrome, magnesium or zinc. This property makes them useful for sorting ferrous (iron-based) metals from non-ferrous ones.

Iron is a 'soft' magnetic material. This means that it is easy to **magnetise** but quickly loses its magnetism. The diagram shows how you can magnetise an iron nail with a battery and some insulated wire.

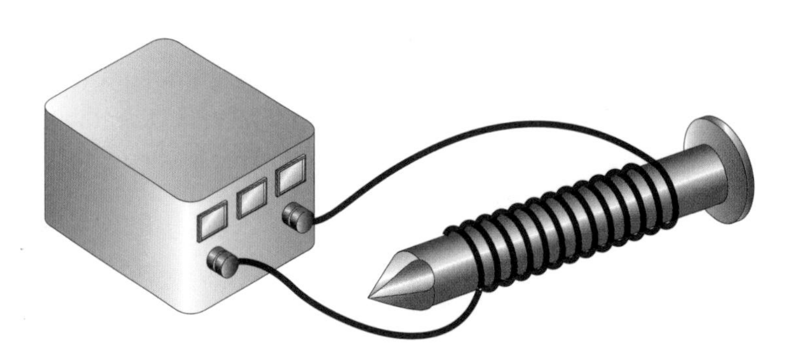

Fig. 8.4 Magnetising an iron nail.

Steel is used to make some permanent magnets.

Steel is harder to magnetise but it keeps its magnetism for a long time. If you test 'copper' coins with a magnet you will find that those minted since 1992 are strongly magnetic because they are made from steel with a thin copper coating.

Magnets have been used for centuries to navigate.

- A magnet that is free to turn round always points with one **pole** (see below) towards magnetic north.
- This is called the **north-seeking pole** or north pole of the magnet.
- The opposite pole is called the **south-seeking** or south pole.

A magnet is better than the Pole Star for navigation – a magnet always points north but the Pole Star is not always visible.

Magnets can attract and repel other magnets. The strongest parts of a magnet are called the poles. Bar magnets have poles at the ends but slab magnets have them along the sides.

When answering questions about attractive and repulsive magnetic forces, always give the reason in terms of opposite or similar poles.

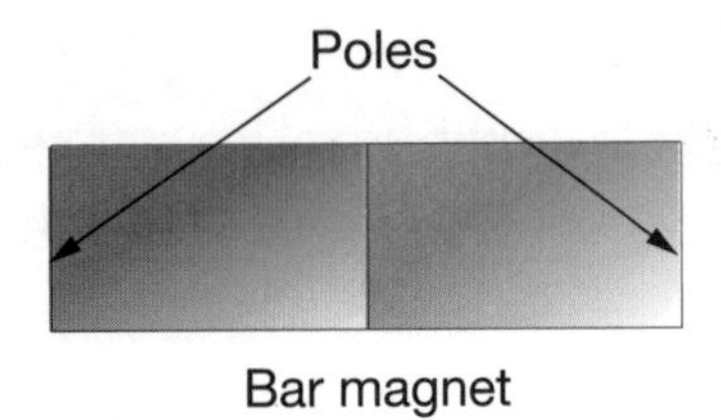

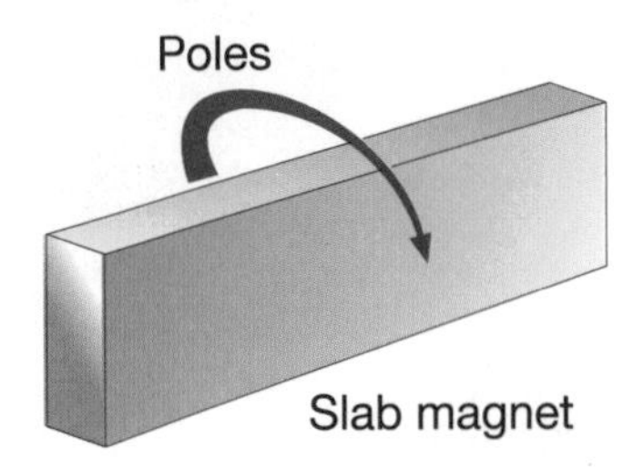

Fig. 8.5

Opposite poles attract and similar poles repel.

This means that the north-seeking pole of a magnet has the opposite magnetism to the Earth's north pole.

1 How could you use a magnet to tell the difference between a brass object and an iron object that has been painted the colour of brass?

2 The north-seeking pole of a magnet points towards the Earth's south pole. True or false?

1. The iron object is attracted to a magnet but the brass one is not attracted. 2. False

Magnetic fields

Always keep a compass well away from any coins that you might have in your pocket or purse.

Compasses use small magnets pivoted on needles to enable them to turn round so that the poles point towards magnetic north and south. However, if a compass is placed near a magnetic material such as iron, it will point towards that instead. When used for navigating, compasses should always be kept away from anything made from iron, nickel or steel.

A compass is a useful tool for investigating the attractive and repulsive forces around a magnet. By looking at the direction of the force on the north-seeking pole of the compass, a set of straight lines can be drawn showing the force pattern all around the magnet. When these lines are joined up they make a pattern.

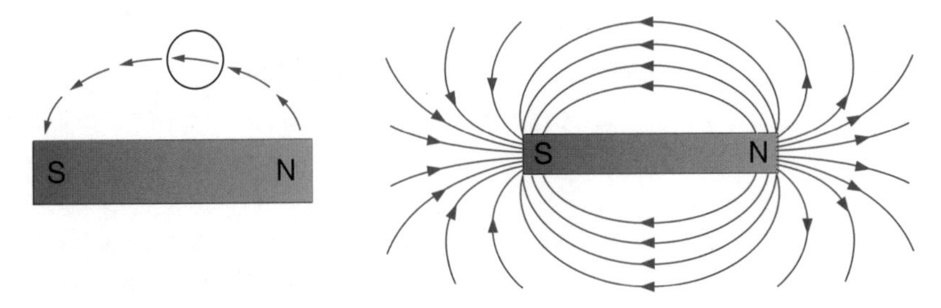

Fig. 8.6 Magnetic forces make a magnetic field pattern.

This is called a **magnetic field pattern** because it shows the pattern of the forces in the magnetic field, which is the region round the magnet where it pushes and pulls magnetic objects. Although the lines are curved, the force at any point is in a straight line.

Key Point

Magnetic field patterns are always drawn to show the direction of the force on the north-seeking pole of another magnet.

It is important to remember that the arrows on a magnetic field pattern always point away from a north-seeking pole and towards a south-seeking pole.

Iron filings act like small compasses – if you sprinkle iron filings around a magnet they line up to show the shape of the magnetic field pattern. They give a 'quick' way of investigating the shape of a magnetic field, but you need to use a compass to see the direction of the forces.

The diagrams show the results of using iron filings to investigate the attractive and repulsive fields between pairs of magnets.

Can you work out which way the arrows go on these magnetic field patterns?

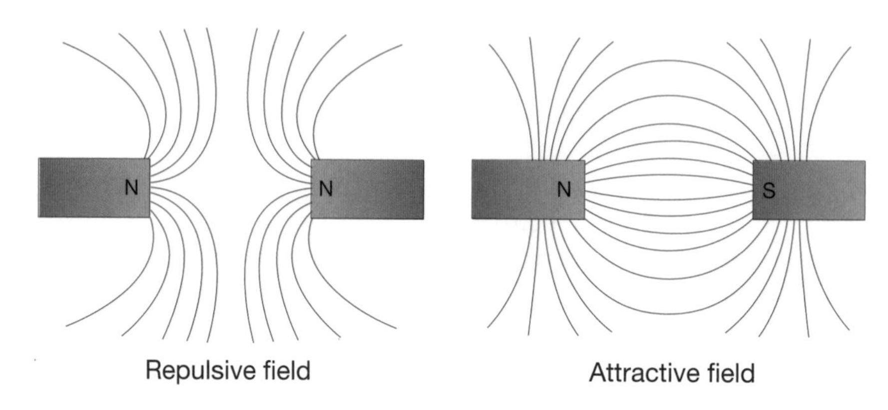

Fig. 8.7

Progress Check

1 Complete the sentence:

The arrows on a magnetic field pattern show the direction of the force on the ___________ -seeking pole of another magnet.

2 A pupil investigates the magnetic field pattern around the end of a steel bar. The diagram shows the results.

(a) What does this tell you about the steel bar?

(b) What would you expect the field pattern to look like around the opposite end of the bar? Explain why.

1. North 2. (a) It is magnetised.
(b) The arrows would point towards the bar. It is the opposite magnetic pole.

Electromagnetism

Electromagnets are all around us. They are used:

- in motors to turn the washing machine drum and the lawn mower blades
- to make a loudspeaker cone vibrate
- to 'write' information in magnetic form onto audio and video tapes and computer discs

All **electric currents** have their own **magnetic field**; this means there are magnetic forces around them. We do not normally notice the effects of these forces because they are very weak. Careful observation using iron filings and a compass shows these field patterns for the current in a **straight wire** and in a **coil**.

Fig. 8.8

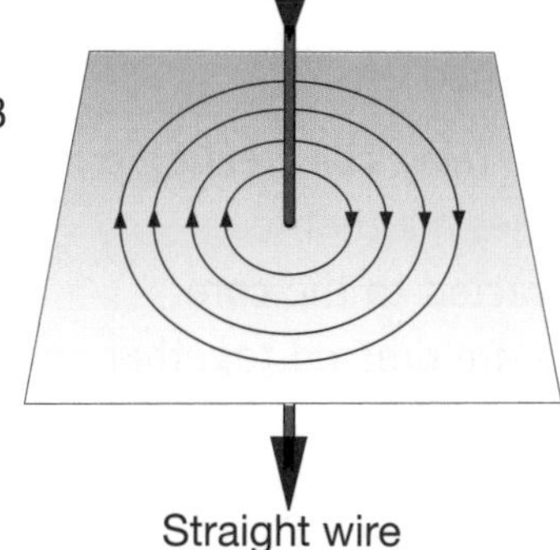

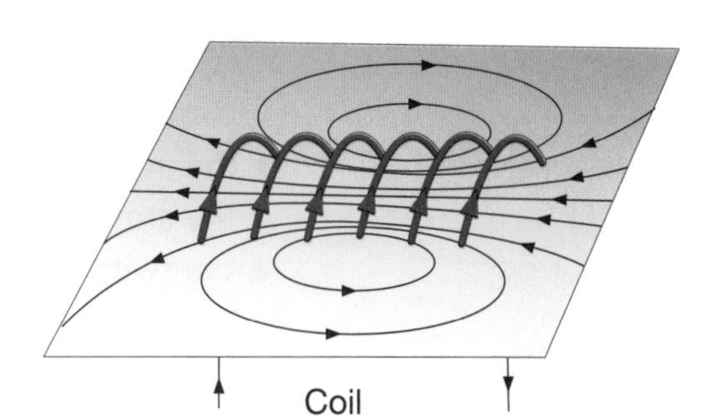

The pattern of the magnetic field around a coil of wire is similar to that of a bar magnet.

Questions about changing the strength of an electromagnet are common in tests and examinations.

Electromagnetic forces can be made much bigger by wrapping a coil of wire around an iron core. When a current passes in the wire, the iron becomes strongly magnetised. This makes a useful electromagnet because, not only can it be switched on and off, its strength can be varied by controlling the current passing in the coil.

Progress Check

1. Give **two** ways in which the strength of an electromagnet can be increased.
2. All copper wires have a magnetic field around them. True or false?
3. Which of these is attracted to an electromagnet?

copper wire **iron nail** **plastic pen** **steel rod**

1. Increase the current or voltage. Wind more turns on the electromagnet.
2. False 3. Iron nail and steel rod.

Using electromagnetism

Electromagnets are also used to sort scrap metals and move iron and steel.

A useful electromagnet is made using a coil of wire on an iron core. This can cause movement when the magnet is switched on and off. Motors, bells and relays all use the movement caused by forces due to electromagnetism.

A **relay** is a switch operated by turning an electromagnet on and off. It may seem strange to use a switch to operate a switch, but using a relay does have advantages in certain situations.

The diagram shows a relay being used to switch a mains lamp.

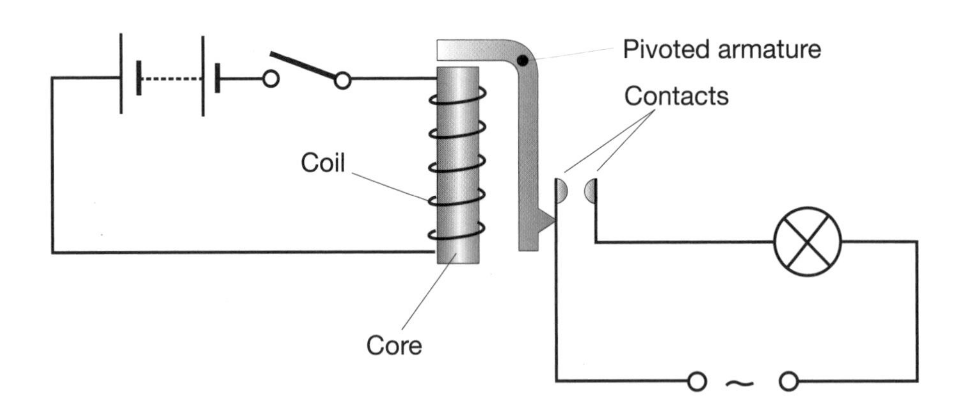

Fig. 8.9

When the relay is switched on:

- current passes in the relay coil creating a magnetic field
- the iron core becomes magnetised
- the armature is attracted to the core
- the switch contacts are pressed together and the light comes on

A relay:

- can operate from a low voltage source and be used to switch devices that work from a higher voltage
- only needs a small current to operate it

The electrical circuits in cars carry large currents that need thick wires. Relays are used so that thin wires can be used for the switches that the driver operates.

Progress Check

1 State **two** uses of electromagnets.

2 Complete the sentence:

A relay uses a small current or voltage to switch on a ____________ current or voltage.

3 Which is the best material for the core of an electromagnet?
Choose from the list.

brass **copper** **iron** **plastic** **steel**

1. Any two from: motors, loudspeakers, bells, relays, lifting iron and steel, sorting iron and steel from other metals. 2. large 3. iron

8.3 Light

After studying this topic you should be able to:

- **understand how light travels and how objects are seen**
- **describe how light is reflected at a mirror**
- **recall the change in direction when light changes speed at a boundary**
- **explain dispersion**
- **work out the colours that objects appear in different coloured lights**

How light travels

Sharp shadows are formed when **light** from a small source, such as a torch, passes around an object that is **opaque**, i.e. that does not let light go through it.

Key Point

These shadows provide evidence that light **travels in straight lines.**

If a low-flying aircraft passes overhead, when you look at where the sound is coming from you see that the aircraft has already gone past. This is because of the different speeds of light and sound.

Key Point

Light travels very much faster than sound.

Typical speeds, in air, are:

- sound – 330 m/s
- light – 300 000 000 m/s

Because of the very fast speed of light, there is normally no noticeable time delay between an event happening and us seeing it. Exceptions to this occur when the distances are vast, for example when looking at stars. Observations of our Sun see what happened eight and a half minutes ago. Light from the second nearest star to us, Proxima Centauri, takes more than four years to get here, so when you are star-gazing you are looking back in time!

Never look directly at the Sun, as it could damage your eyes.

Seeing

Some people think that we see with our eyes, but we also need a brain to be able to see things!

- Eyes are sensors that send messages to the brain along the optic nerve in the form of electrical signals.
- When the brain interprets the signals from the eyes, it assumes that the light travelled to the eyes in a straight line.

- This enables the brain to work out where things are.
- The brain needs light from both eyes to be able to pinpoint the position of objects accurately, as the diagram shows.

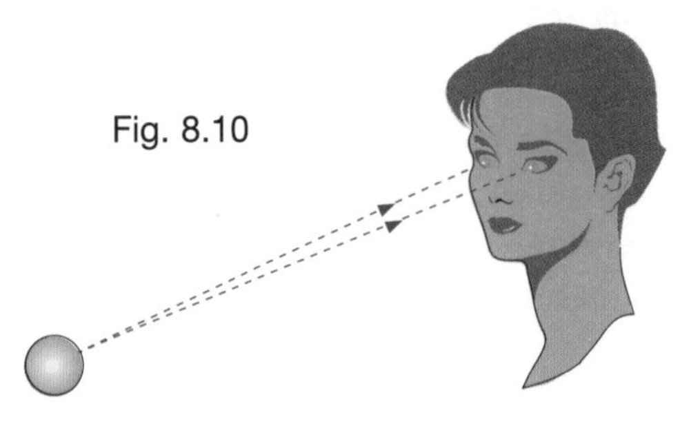

Fig. 8.10

The difference between how we see luminous and non-luminous objects is common in test and examination questions at Key Stage 3.

Television and computer screens are **luminous**: they give out light that our eyes detect. Lamps and the Sun also give out light, as do fires and candles.

We use light sources such as the Sun and artificial lighting so that we can see other things. Most surfaces **scatter** light, that is they **reflect** it in all directions, so that some light enters your eyes even when you move to a different position.

Progress Check

1 Complete the sentence:

A luminous object gives out ________, but a non-luminous object only __________ it.

2 We see an event before we hear the sound from it because light travels faster than sound. True or false?

3 You are sitting in a chair after dark, using a reading lamp to help you to read a book. Describe how you can see the book.

1. light; reflects 2. True
3. Light from the book is scattered (reflected in all directions). Some of it passes into the eyes.

Mirrors

Unlike most everyday things, **mirrors** do not reflect light in all directions but they do reflect it in a regular and predictable way.

Key Point

Light is reflected from a mirror at the same angle as it hits it.

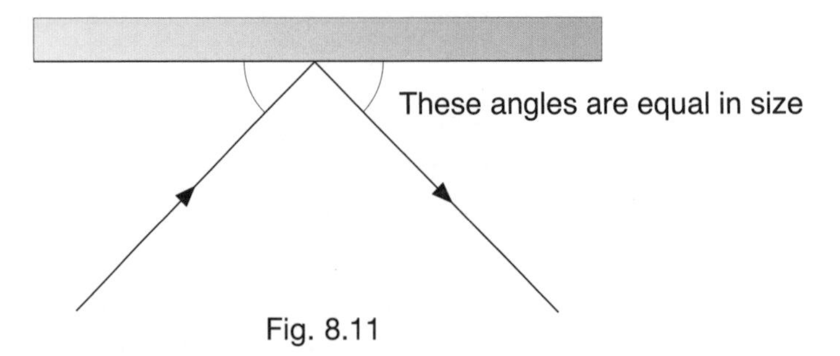

Fig. 8.11

A common misconception is that the image in a mirror is 'laterally inverted' or turned round sideways. It isn't. When you look into a mirror, you see the image of the left side of your face on your left.

We use this regular reflection when we look at our image in a mirror. Looking into a mirror causes our brain to get confused: it 'sees' things that aren't really there!

- Light from your nose (and other parts of you!) hits the mirror and is reflected at equal angles. The reflected light is detected by your eyes.
- Your brain then 'sees' the nose, and works out where it is, assuming that the light has travelled in straight lines.
- In this case it gets things wrong: it 'sees' a nose behind the mirror.

The dotted lines show where light does not actually travel, but only appears to come from.

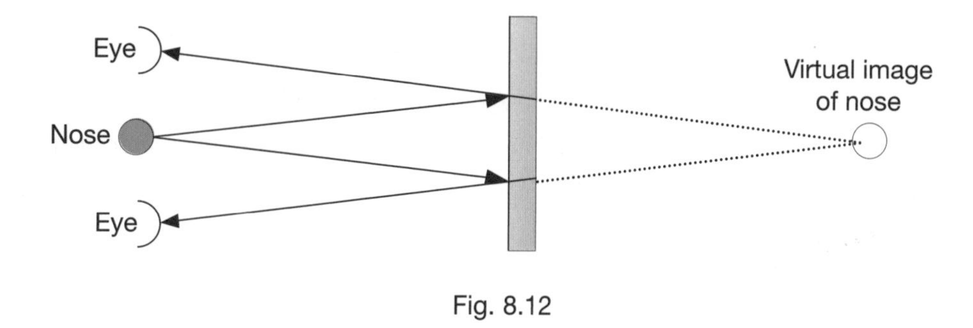

Fig. 8.12

What you see is called a **virtual image**. The word 'image' just means 'likeness', and in all respects the image is a likeness of the real thing; it is the same way up and the same size and colour. The image is called virtual because, unlike the image that you see on a television or cinema screen, it is not really there.

Using the rule about the way in which light is reflected at a mirror, you can construct images. If you do this, you will see that:

The image in a mirror is always formed straight behind the mirror, the same distance behind it as the object is in front of the mirror.

Many pupils at this level think that the image is actually on the mirror, rather than behind it.

Mirrors are used in periscopes to turn light round corners. Light hitting a mirror at an angle of 45° is reflected at the same angle and so is effectively turned through 90°. The diagram shows how two mirrors are arranged inside a periscope. You can make a periscope quite easily using two small mirror tiles and the cardboard tube from the inside of a roll of kitchen foil.

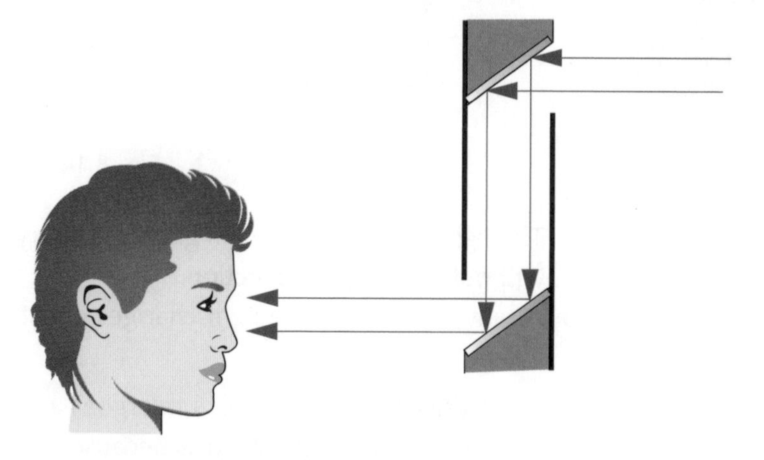

Fig. 8.13 Two mirrors are arranged to make a periscope.

Progress Check

1 Which diagram shows the reflection of light by a mirror correctly?

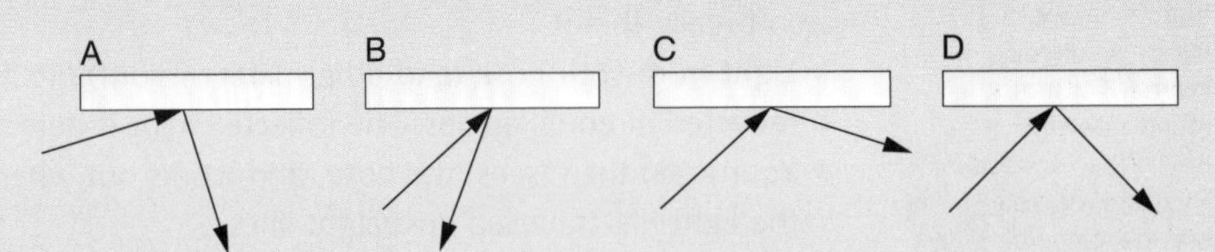

2 When you look in a mirror, is the image that you see real or virtual?

3 A The image in a mirror is further from the mirror than the object is.

B The image in a mirror is closer to the mirror than the object is.

C The image in a mirror is the same distance away as the object is.

Which is correct – A, B or C?

1. D 2. Virtual 3. C

More images

Have you ever noticed how the water in a swimming pool never looks to be as deep as it says it is? This is another example of your eye-brain system being fooled when light does not travel in straight lines. A **change in the speed** at which light travels can also cause a **change in direction**.

As light enters water or any other dense transparent material it slows down.

Typical speeds of light in water and glass are:

- water – 230 000 000 m/s
- glass – 200 000 000 m/s

The change in speed of light as it enters water or glass is called **refraction**. The diagrams show the effect of the change in speed when light passes through a sheet of glass.

The greater the angle at which the light meets the air-glass boundary, the greater the change in direction.

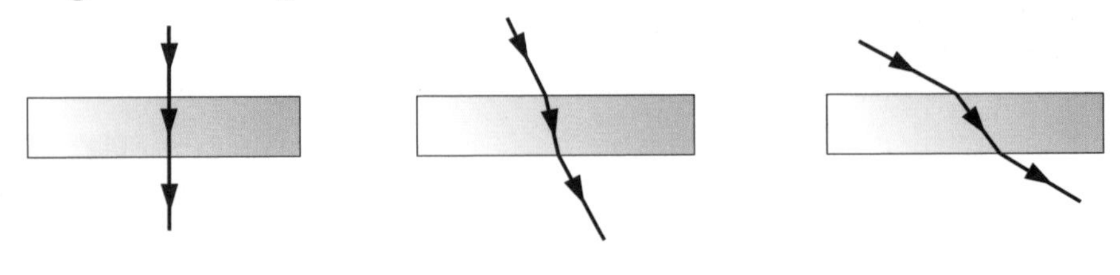

Fig. 8.14

- The light that meets the air/glass boundary at an angle of 90° carries on without a change in direction.
- At any other angle the light changes direction as it goes into and leaves the glass block.

Light travelling through water undergoes a similar change in direction. The amount of change is slightly less because there is less change in speed as light enters water.

> When viewed through water, objects only appear to about three-quarters of their real distance away.

So why does the swimming pool look to be shallower than it really is? The answer is because of the change in direction that occurs when light leaves the water and speeds up as it enters the air.

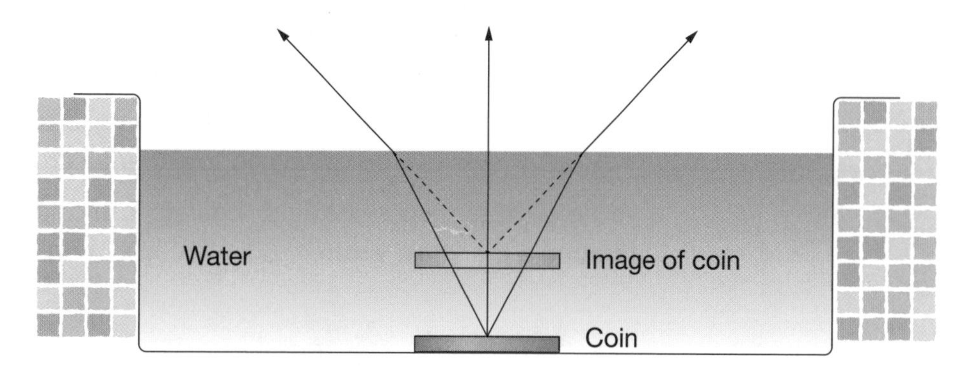

Fig. 8.15 If you drop a coin into a swimming pool, the image appears closer than it really is.

The dotted lines in the diagram show where the light appears to have come from, assuming that it has travelled in straight lines. This is another example of a **virtual image**. As with a mirror, the virtual image is just like the original object, except that it is not really there.

Progress Check

1. Complete the sentence:
   When light passes from air into glass it slows down. This effect is called ________________.
2. The bottom of a swimming pool appears to be closer than it really is. True or false?
3. The diagram shows light from a lamp passing through a block of glass.

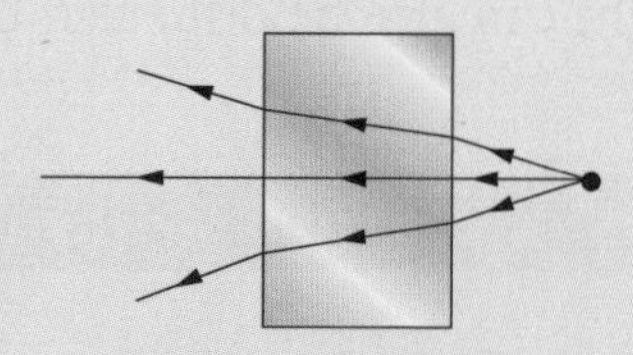

When a person looks at the lamp through the block of glass, the lamp looks to be nearer than it really is.

(a) Draw on the diagram to show where the lamp looks to be.
(b) What have you assumed about the way in which light travels?

1. refraction 2. True 3. (a) The diagram should show the light leaving the block traced back using straight lines; the point where these lines cross is where the lamp looks to be. (b) That light travels in straight lines.

Colour

> When studying this section, it is important that you understand the difference between colour addition and colour subtraction.

Manufacturers of paint make different **colours** and shades by starting with a white paint and then adding dyes to it. Until the time of Isaac Newton, it was thought that coloured objects did the same thing to light. Newton realised that white light already contains all the colours and, when light is reflected by a coloured object, colours are not added but taken away.

Coloured objects do not add colour to white light; they take it away.

Newton noticed that:

- white light going into a glass **prism** emerged split into different colours
- these colours can be combined together again to make white

You may have seen white light separate into colours when it passes through a triangular prism.

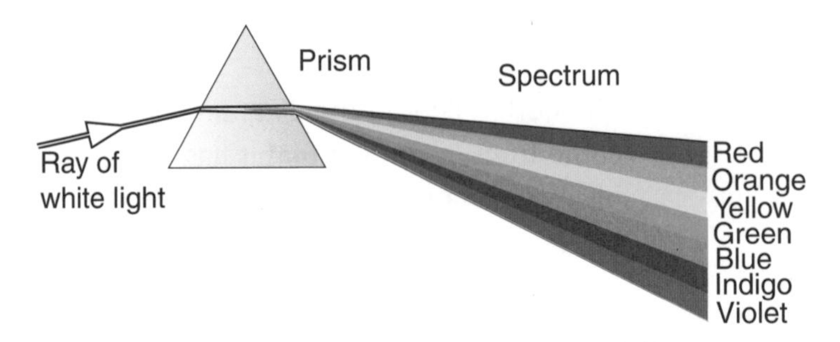

Fig. 8.16

The separation of light into colours is called dispersion. It is caused by the different colours of light travelling at different speeds in the prism.

Light can also be dispersed by water droplets, forming a **rainbow**.

Colour addition

Colour addition occurs when different coloured lights shine on, or from, a screen.

When you watch television you are seeing the effects of combining colours together. A television screen can glow with three different colours: **red**, **green** and **blue**. These are called the **primary** colours because all other colours can be made by combining these three colours.

The diagram shows how the three **secondary** colours, **yellow**, **magenta** and **cyan** (turquoise) are each produced by **adding together** the light from two primary colours.

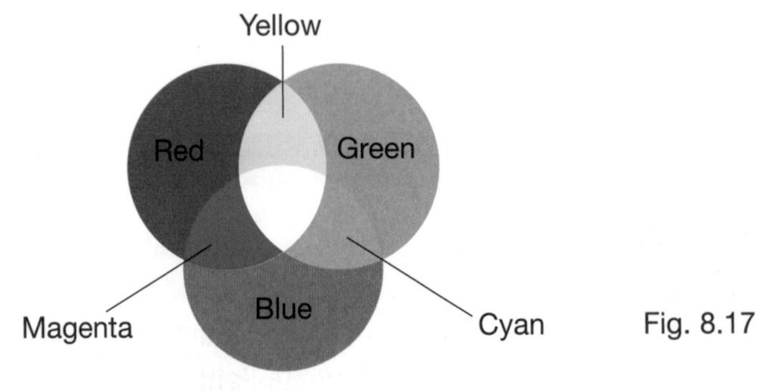

Fig. 8.17

Televisions reproduce colour by using the three primary colours, but pictures in magazines use the three secondary colours.

The area in the middle, where all three primary coloured lights add together, is white.

Progress Check

1 Write down which two primary colours, from red, green and blue, are added together to make:

(a) yellow
(b) magenta
(c) cyan

2 What causes light to be dispersed by a glass prism?

3 Complete the sentence:

When light changes direction as it enters a prism, __________ light has the smallest change.

1. (a) Red and green (b) Red and blue (c) Green and blue 2. Different colours of light travel at different speeds in the prism and change direction by different amounts. 3. red

Colour subtraction

Televisions and computer screens can reproduce any colour by adding the three primary colours together in varying proportions. The colour that non-luminous objects appear depends on the colours that they **absorb**.

Key Point

Coloured objects and colour filters absorb or take colours out of the light that falls on them.

- **Primary** colours take out everything except their own colours.
- Each **secondary** colour removes just one primary colour; the colour removed is the one that isn't used when mixing the secondary colour from red, blue and green light.
- So cyan removes red light, yellow removes blue light and magenta takes away green light.

If you go to a disco where there are different colours of light, you will notice that clothes seem to take on different colours. A cyan-coloured T-shirt looks black in red lighting and green in yellow lighting.

To work out what colour different objects look under coloured lights:

- start with the primary colours present in the light
- take away the colours that the object absorbs
- you are left with the colours that are reflected

It is important to remember that an object that appears black does not reflect any light.

As you can see, it's just a simple subtraction sum.
Sometimes an object takes away all the colours in the light; when this happens it appears black.

Progress Check

1 A boy wears a sweater which is turquoise (cyan) when viewed in daylight.

(a) Which primary colour is absorbed by the turquoise sweater?

(b) What colour would the sweater look to be when viewed in the following colours of light?

(i) yellow

(ii) red

(iii) blue

(iv) magenta

2 White light passes through a yellow filter and then a cyan filter. What colour emerges from the cyan filter?

1. (a) Red (b)(i) Green (ii) Black (iii) Blue (iv) Blue 2. Green

8.4 Sound and hearing

After studying this topic you should be able to:

- **describe how sound is produced**
- **explain how changing the amplitude and frequency of a sound changes the sound that is heard**
- **describe how sound is detected by an ear**
- **explain why ears need to be protected from loud sounds and how this is done**

Sound production

When answering questions about what causes sound, always stress that sound is caused by an object vibrating.

To make a **sound**, something has to **vibrate** in a to-and-fro motion. Televisions, radios and hi-fi all use **loudspeakers** to reproduce sound. A loudspeaker consists of a paper cone driven backwards and forwards by an electromagnet.

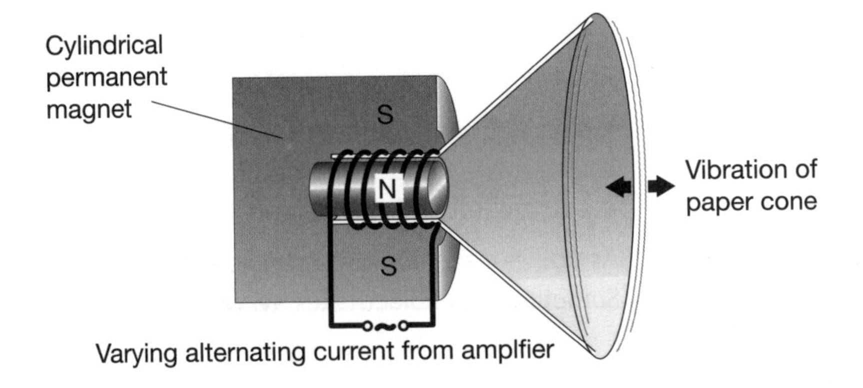

Fig. 8.18 A loudspeaker.

Some musical instruments have strings that vibrate when they are hit or plucked; others have columns of air that vibrate when they are blown.

Travelling sound

Sound travels slowest in gases because the particles are the most widespread.

Sound can travel through anything that has particles capable of transmitting the vibrations. It travels very fast in solids, where the particles are close together, slower in liquids and slowest of all in gases.

Key Point **Sound cannot travel in a vacuum – it needs a material to transmit the sound by vibrations of its particles.**

You may have seen a demonstration of a bell ringing inside a glass jar. Sound from the bell travels through the air and the glass and then air again to the ear. When air is removed from the jar, the sound cannot be heard.

It is important to remember that sounds can only be transmitted by the vibration of particles.

A **slinky spring** can be used to show how sound waves travel through materials. As the wave moves along the spring, each part of the spring vibrates. The particles in air vibrate in a similar way when they transmit sound. Each wave consists of a **compression** or squash, where the particles are close together, followed by a **rarefaction** or stretch, where the particles are further apart.

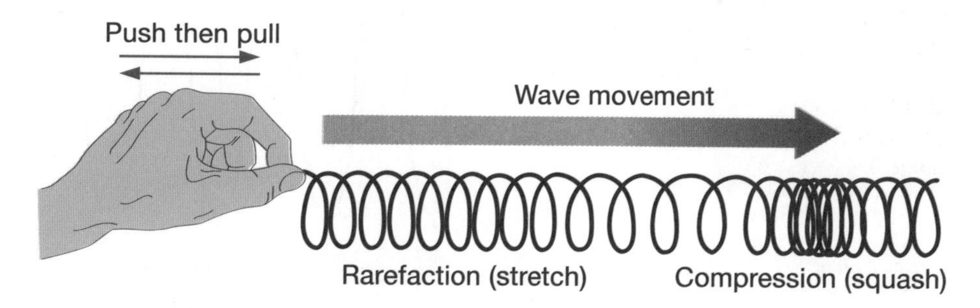

Fig. 8.19

A hand pushing and pulling a slinky spring is a good model of what happens when a loudspeaker cone is pushing and pulling on the air.

- Using this model, **moving the hand further** has the same effect on the spring as **turning up the volume control** has on the air particles.
- The particles **move further** and you hear a **louder** sound.
- When you push further you are increasing the **amplitude** of the vibration.

Key Point **The amplitude of the vibration is the greatest distance that each part of the slinky moves from its rest position.**

Pupils are often confused by this – they see an up-and-down movement on the screen that actually represents a to-and-fro movement.

A microphone and an oscilloscope can be used to plot a graph that shows the movement of air particles when a sound wave passes. The upwards and downwards movement of the oscilloscope trace represents the forwards and backwards movement of an air particle.

The distance marked with an arrow (Fig. 8.20) represents the amplitude of the wave. A louder sound of the **same pitch** has a larger amplitude.

When asked to label the amplitude of a wave on a displacement-time graph, many pupils label the distance between the maximum displacements in each direction. This is wrong – the amplitude is only half this distance.

The trace on an oscilloscope screen shows the number of sound waves that are detected in a certain time; the actual time can be changed by adjusting the oscilloscope settings. A 'crest' and a 'trough' show one complete wave. The trace shown in the diagram below shows two and a half waves.

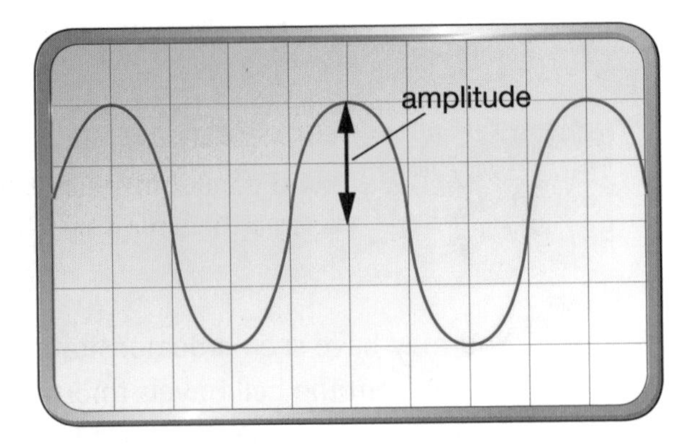

Fig. 8.20 An oscilloscope trace of two and a half waves.

Make sure that you know how amplitude and frequency are shown by the trace on an oscilloscope screen. Questions about this are common at Key Stage 3.

The next diagram shows the trace when a note of higher pitch is played.

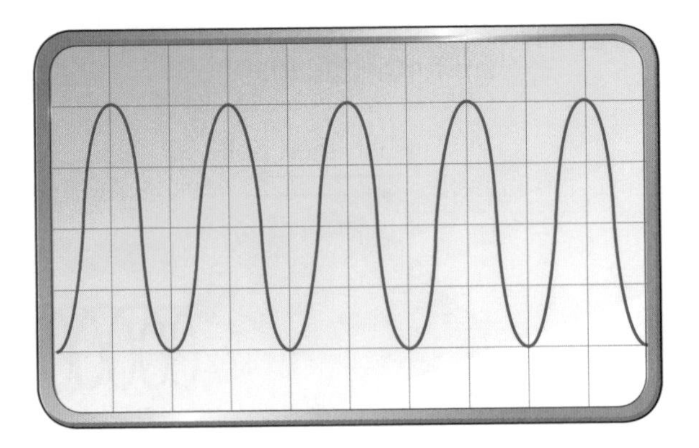

Fig. 8.21 A trace of a higher pitched note.

There are now more waves in the same time; the **frequency** (the number of waves per second) has increased.

Key Point

Frequency is measured in hertz (Hz). 1 Hz = 1 wave per second.

These measurements using an oscilloscope display show that:

- increasing the **amplitude** of a sound wave makes it sound louder
- increasing the **frequency** of a sound wave gives it a higher pitch

Progress Check

1 Complete the sentence:

All sounds are caused by ______________.

2 Sound can only travel in solids. True or false?

3 Choose the correct word from the list to complete the sentence.

Frequency is measured in ______________.

amps **hertz** **metres** **seconds**

4 You can make a sound by plucking a stretched rubber band.

(a) How does the rubber band make a sound when it is plucked?

(b) Describe how the sound is carried from the rubber band to your ears.

(c) How does the sound change if the rubber band is plucked harder? What causes this change?

(d) If the rubber band is stretched more, when it is plucked the sound has a higher pitch. What change has caused the higher pitch?

1. Vibrations. 2. False. 3. Hertz. 4. (a) By vibrating. (b) By vibrations of the air particles. (c) It sounds louder. The vibrations have a greater amplitude. (d) There are more vibrations each second or the frequency has increased.

Hearing

Sound reaches our ears as the energy of vibrating air particles. The vibrations have a tiny amplitude, typically a few thousandths of a millimetre, but our ears are very sensitive to small amounts of energy transmitted in this way.

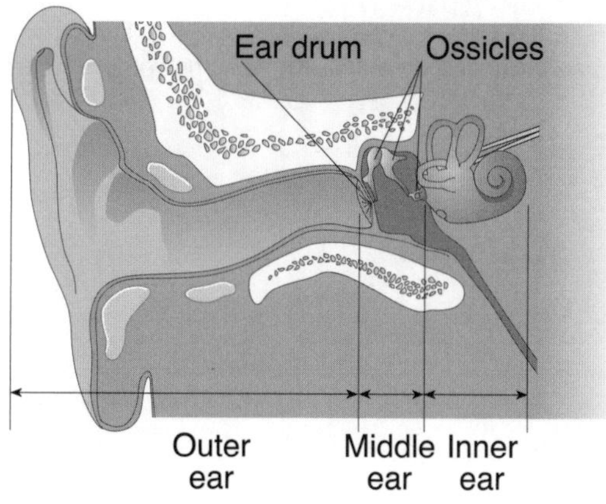

Fig. 8.22 The ear.

In an ear:

- sound causes vibrations of the ear drum in the outer ear
- the ossicles are three bones in the middle ear that transfer this energy to the inner ear
- **nerve endings** in the inner ear are stimulated by the vibration and send messages along the auditory nerve to the brain. These messages are sent as electrical impulses.

Not everybody can hear the same sounds.

- The ears of a normal young person detect sounds with a frequency ranging from about 20 Hz to about 20 000 Hz.
- As you get older, the range of frequencies that you can hear is reduced.
- A middle-aged person may not be able to hear sounds with a frequency greater than 15 000 Hz although the hearing at low frequencies is less likely to be affected.
- Even within this range, your ears are more sensitive to some frequencies than to others, sounds at a frequency of around 2000 Hz sounding louder than higher or lower sounds.

Small animals communicate using higher frequencies than those used by large animals.

Other animals have different frequency ranges. Bats can emit and detect sound waves that are well above the range of human hearing, and dog whistles use a high frequency that humans cannot hear. Elephants and dolphins can communicate over long distances using very low frequencies that have a longer distance range than high frequency waves.

People who work in noisy environments should wear ear muffs to protect their ears from damage. Sudden exposure to a loud sound such as an explosion can cause immediate damage by breaking the ear drum or the ossicles.

The ossicles are also subject to wear; they are pieces of machinery in constant use. A person who is repeatedly subjected to loud sounds, such as those from pneumatic drills or discos, suffers loss of hearing as the ossicles wear away. Surgeons can sometimes replace worn out ossicles with plastic ones to improve a person's hearing.

Progress Check

1. Which two frequencies of sound waves cannot be detected by most humans?

   10 Hz 35 Hz 1000 Hz 15 000 Hz 25 000 Hz
2. Disc jockeys and pop singers often suffer from hearing loss at a young age. Suggest what causes this.
3. What is the name of the three bones in the middle ear?
4. The ear drum can be damaged by poking a sharp object in the ear. True or false?

1. 10 Hz and 25 000 Hz. 2. The loud sounds that they are exposed to cause excessive wear of the ear bones (ossicles). 3. The ossicles. 4. True

Practice test questions

The following questions test levels 3-6

The diagram shows a central heating radiator. The radiator is made of steel and is filled with hot water.

(a) Explain why the radiator is hot to the touch. **[2]**

....

....

(b) **(i)** By what method is the air above the radiator heated? **[1]**

....

(ii) Explain how this happens. **[2]**

....

....

(c) The radiator also radiates energy.

What carries the radiant energy away from the radiator? **[1]**

....

The following questions test levels 5-7

(a) Tara is choosing what to wear to the disco. She has three tops to choose from.

One is white, one is red and the third is yellow. She shines different coloured lights on them to see how they will look.

Complete the table to show the colours that the tops appear to be. **[6]**

Colour in white light	Colour in blue light	Colour in red light
white		
red		
yellow		

(b) Tara looks at herself in the mirror.

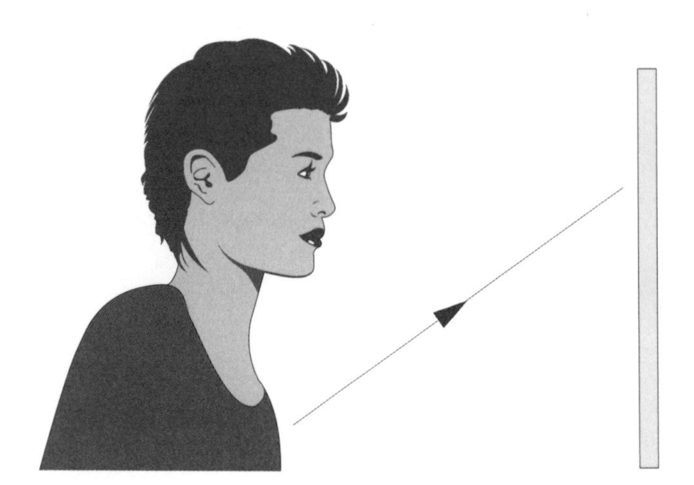

(i) Complete the diagram to show how light from her top is reflected at the mirror. **[1]**

(ii) Describe how Tara's top and the mirror reflect light in different ways. **[2]**

....

....

Year 9

Chapter Nine

The topics covered in this chapter are:

- **Energy and electricity**
- **Gravitation and space**
- **Speeding up**
- **Pressure and moments**

9.1 Energy and electricity

After studying this topic you should be able to:

- **describe energy transfer and energy flow processes**
- **explain how the energy transfer in a circuit depends on the voltage of the source and how voltage is measured**
- **describe how electricity is generated**

Energy transfer

Questions in Key Stage 3 tests often ask you to identify the type of energy that an object has, for example kinetic energy or gravitational potential energy.

Every event involves **energy** and an **energy transfer**. Devices that we use in our everyday living transfer energy. The diagrams show some examples.

Fig. 9.1

The kettle transfers energy from **electricity** to **heat** in the water.

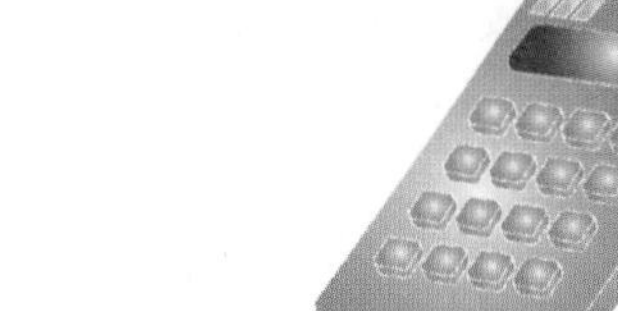

Fig. 9.2

The solar cell in this calculator transfers energy from **light** into **electricity**.

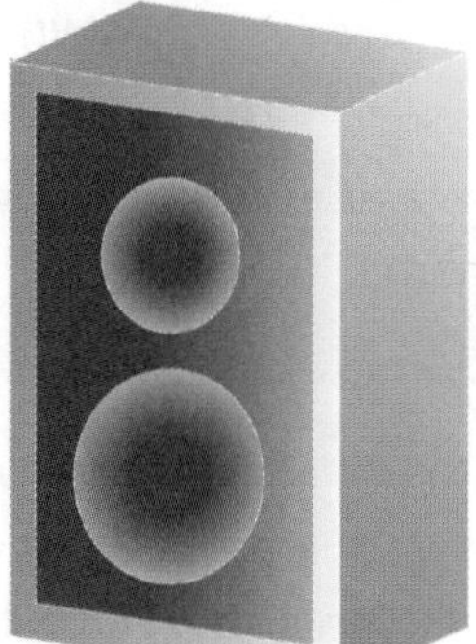

Fig. 9.3

The loudspeaker transfers energy from **electricity** to **sound**, which is **kinetic energy** of the air particles.

Gravitational potential energy is sometimes referred to as just 'potential energy'.

Fig. 9.4

The ski lift transfers energy from **electricity** to **gravitational potential energy** of the people as they are lifted up.

Fig. 9.5

As the bus speeds up, **chemical energy** stored in the fuel is transferred to **movement** or **kinetic energy** and **heat** in the exhaust gases.

Progress Check

1. Use words from this list to complete the sentence.

   chemical gravitational potential heat kinetic

   When a weightlifter lifts some weights, energy is transferred from ____________ energy in the weightlifter's cells to _____________ energy in the weights.
2. A moving bus has more kinetic energy than one that is stationary. True or false?
3. What energy transfer takes place when coal burns on a fire?

1. Chemical, gravitational potential. 2. True 3. Chemical energy to heat and light.

Energy flow

The lamp is not gaining any energy because it operates at a constant temperature.

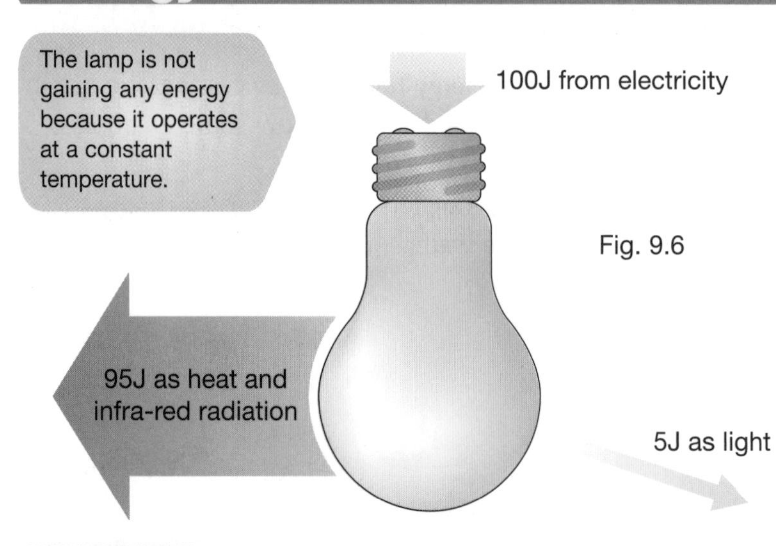

Fig. 9.6

When a kettle is used to heat some water, the kettle and its contents are gaining energy as they warm up. When a lamp is used to light a room, energy **flows** through the lamp at a constant rate. The diagram shows the energy flow through the lamp each second.

All the energy that flows into the lamp also flows out – this is known as **conservation of energy**.

Do not confuse this with 'conserving energy resources', which means taking action to make our fossil fuel supplies last longer.

Key Point

Energy is never created or destroyed – this is the principle of conservation of energy.

If you turn off a light at the mains switch, the room goes dark immediately. Light doesn't stay as light for very long after leaving the lamp, so the light in a room has to be continually replaced. What happens to the energy that leaves the lamp?

- The hot lamp loses energy to the surrounding air – this is carried away by a convection current.
- The light and infra-red radiation are absorbed by the walls and other surfaces – causing them to warm up.

The temperature rise is so small, it would be very difficult to detect with a thermometer.

The effect is that all the energy from the lamp is spread out or **dissipated** in the room, causing a very small temperature rise.

Almost all the energy that we take from sources such as electricity, gas, coal and petrol ends up as heat in our surroundings – in the buildings that we live in, the air and the outdoors. We cannot get this energy back very easily; it is much easier to obtain more energy from a fuel or electricity than to extract the energy from the air and the ground outside.

Progress Check

1. Explain why it is not possible to obtain more energy out of an appliance than the amount that goes in.
2. Complete the sentence:
   When energy spreads out it is said to be ________________.
3. Almost all the energy transferred from electricity ends up as heat. True or false?

1. Energy is always conserved – it cannot be made. 2. dissipated 3. True

Energy in circuits

In a rechargeable battery the chemical changes can be reversed.

A torch needs to be portable, so it uses batteries for its energy supply. Milk floats also use batteries which are recharged from the electricity mains supply overnight. When a torch is being used:

- chemical changes take place in the battery
- these changes release energy
- the electric current transfers the energy to the lamp

A milk float needs much more energy than a torch does, so it uses batteries that have a higher **voltage**.

Measuring voltage

Batteries can have different voltages and laboratory power supplies usually have different voltage settings.

The diagrams show how a voltmeter is used to measure the voltage of a power supply and across one of the lamps in a series circuit.

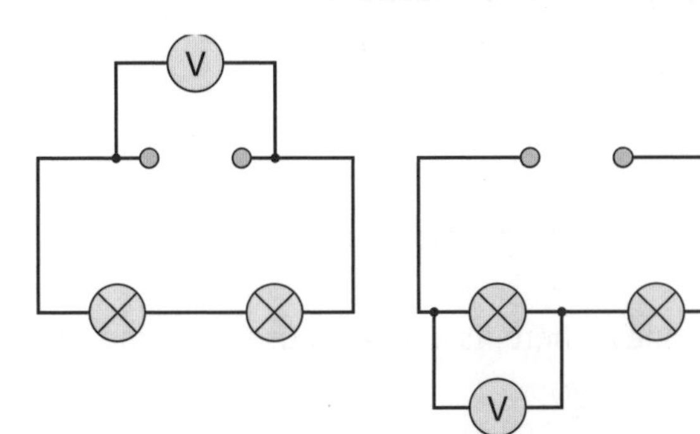

This voltmeter measures the voltage of the power supply.

This voltmeter measures the voltage across the lamp.

Fig. 9.7

Key Point

A voltmeter is a device that measures voltage in volts (V). It is always connected in parallel with a power supply or component.

If you connect a voltmeter in series in a circuit, the circuit will not work because a voltmeter has a very high resistance.

Voltage is a measure of the energy transfer to and from the charged particles that carry the current in a circuit. Increasing the voltage in the circuit shown above increases the brightness of the lamps. This is because:

- increasing the voltage increases the current
- the moving charged particles have more energy to transfer from the power supply to the lamps

Progress Check

1 Complete the sentence:
The source of energy in a battery is ________________ energy.

2 To measure the voltage across a lamp, should a voltmeter be connected in series or in parallel with it?

3 What two factors determine how bright a lamp appears in a circuit?

1. chemical 2. In parallel. 3. The current and the voltage.

Generating electricity

Electricity cannot be stored, so it has to be generated 'on demand'. Most of our energy is generated using fossil fuels, but some is generated from nuclear power and a small amount from renewable energy resources such as wind and solar energy.

The diagram shows how a coal-burning power station produces electricity.

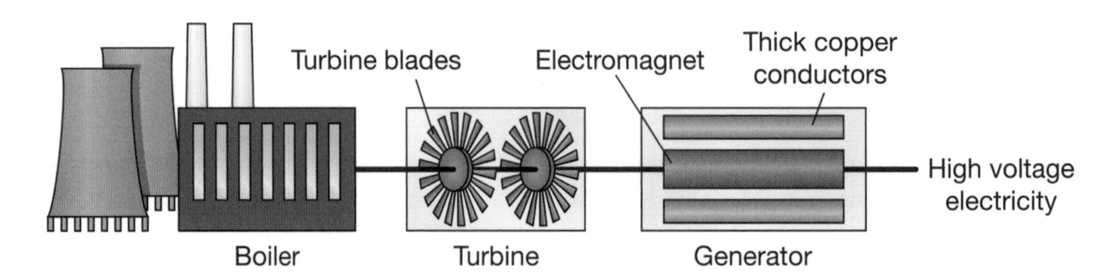

Fig. 9.8

A common error is to state that energy from the steam is transferred to kinetic energy of the turbines. But the turbines do not get faster or slower. There is a constant flow of energy through the turbines and generator.

- Coal is burned in the boiler.
- Energy from the burning coal is used to generate steam at a high temperature and pressure.
- The steam is piped to the turbines where its energy is used to keep the blades turning.
- The turbine drives the generator.

Inside the generator an electromagnet spins inside thick copper conductors. The movement of the magnetic field causes a voltage and a current in the copper conductors – this is how electricity is generated.

Nuclear power stations operate in a similar way.

- Nuclear fuel is contained in a reactor.
- Energy is released when large atomic nuclei are split into smaller ones.
- The energy is removed by a coolant and used to keep turbines spinning.

Progress Check

1. At night energy from electricity is stored by pumping water from a low reservoir to a high one. What type of energy does the water gain?
2. A nuclear power station does not burn fossil fuel. True or false?
3. Which part of a power station drives the generator?
4. What other substance is needed in a boiler to extract the energy from coal?

1. Gravitational potential energy. 2. True 3. The turbine. 4. Air or oxygen.

9.2 Gravitation and space

After studying this topic you should be able to:

- **explain that gravitational forces affect all masses**
- **describe how the Sun's gravitational pull on a planet depends on its mass and distance from the Sun**
- **state some uses of artificial satellites**

Gravitational force

Gravitational forces are always attractive – they can only pull, not push.

The Moon goes round the Earth and the Earth goes round the Sun. This rotation is due to the **gravitational force** that exists between any two objects. If you tie a rubber bung onto the end of a piece of string you can whirl it round your head in a circle, but it only keeps going round you as long as you are pulling on the string. The motion of objects in the Solar System is due to the gravitational pulling forces between the Sun, the planets and their moons.

Key Point

The Sun's gravitational pull keeps the planets in orbit.

When describing gravitational forces on planets, always use a phrase such as 'the Sun's pull on the Earth'.

The diagram shows the strength of the Sun's gravitational pull at the three innermost planets.

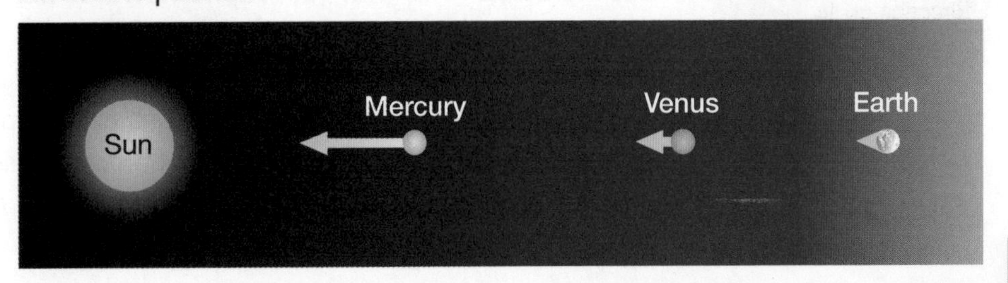

Fig. 9.9

As you can see, the Sun's gravitational pull is weaker at planets further away. The strength of the Sun's gravitational pull determines the speed of a planet in its orbit; the closer a planet is to the Sun, the greater the pull and the faster it moves.

Gravitational forces:

- act between all objects that have mass
- increase in strength with increasing mass
- decrease in strength with increasing separation of the masses

Progress Check

1. Give two reasons why the size of the force pulling Pluto towards the Sun is smaller than the force on any other planet.
2. Suggest why two people who are sat next to each other are not pulled together by gravitational forces.
3. Gravitational forces only push; they do not pull. True or false?

1. Pluto has the least mass of all the planets. It is the furthest planet away from the Sun.
2. The gravitational force between them is tiny and not big enough to overcome the resistive force of friction. 3. False

Planetary orbits

Seasons are caused by the Earth's tilt on its axis – the changing distance from the Sun has little effect.

All the planets move in orbits that are **ellipses**, like circles that have been squashed a bit. For all the planets except Mercury and Pluto the ellipses are very near to being circles so the distance between the Earth and the Sun does not change very much during a year. Mercury's orbit takes it closer to the Sun at some times of its year and this affects its speed.

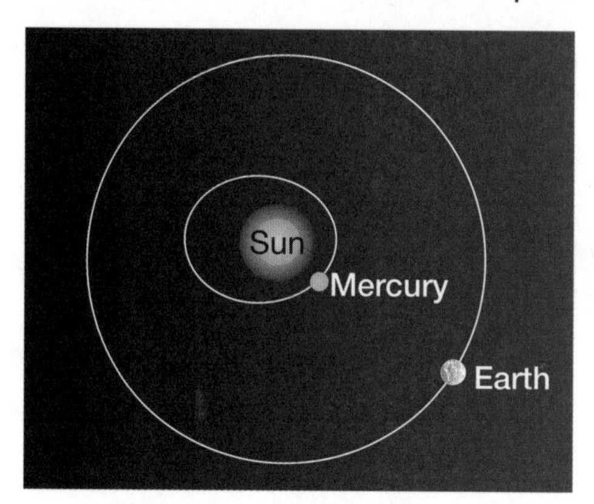

Fig. 9.10 Mercury's orbit takes it closer to the Sun at some times of its year.

Mercury is pulled towards the Sun, so as it approaches the Sun the force on it is in the same direction as its motion.

As Mercury gets closer to the Sun, it speeds up in its orbit and it slows down as it gets further away. Pluto behaves in a similar way.

At Jupiter, the strength of the Sun's pull is even weaker than at the Earth, but the actual force that pulls Jupiter towards the Sun is bigger than that pulling Earth.

This is because Jupiter is more massive than the Earth so there is a lot more matter for the Sun to pull on! Jupiter still moves more slowly than the Earth because the **mass** of a planet does not affect the speed of its orbit; its **distance** from the Sun is the key factor.

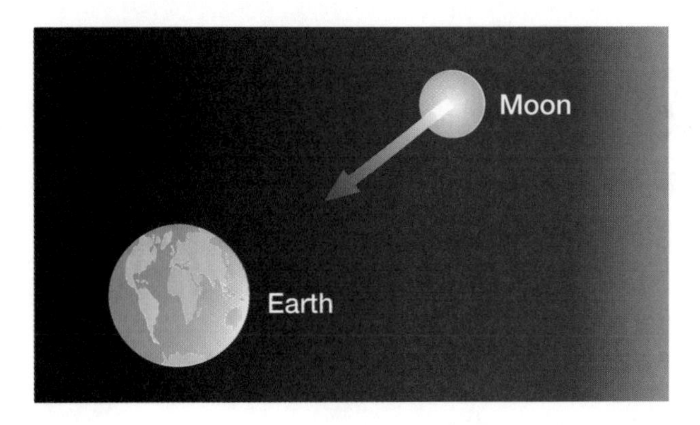

If the Sun's pull keeps the planets going round it, what keeps the Moon going round the Earth? The answer is the Earth's pull. Just as the Earth pulls us towards it, it also pulls the Moon.

The Moon also pulls the Earth – this is the main cause of tides.

Fig. 9.11 The Earth pulls the Moon towards it.

Progress Check

1 What shape are the orbits of planets? Choose the best word from the list.

circles ellipses squares

2 Complete the sentence:
As Mercury approaches the Sun, its speed in its orbit __________.

3 Which planet experiences the greatest pulling force from the Sun? Choose from the list.

Earth Jupiter Mercury Pluto

1. Ellipses 2. increases 3. Jupiter

Satellites

You may have noticed a time delay between the question and answer in cross-Atlantic television interviews. This is because of the time it takes for the waves to travel the vast distances to and from satellites.

The Moon is the Earth's **natural satellite**, but the Earth also has hundreds of artificial satellites that do a wide variety of jobs.
Satellites are used:

- to take **photographs** of the Earth at regular intervals so that weather forecasters can see the changes in **weather patterns**
- for **television broadcasting** and for communicating by telephone and radio between places that are 'hidden' by the Earth's curvature
- by ships, cars and aircraft for **navigating**
- by some countries to **spy** on others
- by astronomers – telescopes such as the **Hubble telescope** have a much better vision in orbit around the Earth, where the light they receive from stars has not been distorted by passing through the Earth's atmosphere

Progress Check

1 Satellites are kept in orbit around the Earth by the Sun's gravitational pull. True or false?

2 Complete the sentence:
Satellites placed in orbit above the Earth's __________ have a better view of space than those on the Earth's surface.

3 Explain why a satellite is needed to send television pictures from the UK to Australia.

1. False 2. atmosphere 3. Television signals travel in straight lines. The curvature of the Earth prevents the signals from travelling there directly.

9.3 Speeding up

After studying this topic you should be able to:

- **compare the speeds of moving objects by comparing the times taken to travel a given distance**
- **calculate the speed of a moving object**
- **use the relationship between speed, distance and time to calculate journey times and distances**
- **interpret graphs that represent motion**

Who won the race?

Sprinters, marathon runners and Grand Prix racing drivers all compete over a fixed distance. The winner is the person who arrives at the finishing post in the shortest time. Measured over the whole race, the person who takes the shortest time has the greatest **speed**.

To work out the speed of a moving object two measurements are needed:

- the distance travelled by the object
- the time taken to travel that distance

Fig. 9.12 The cyclist's average speed can be worked out from measurements of the distance travelled and the time taken.

For example, if a bus travels 60 miles in 2 hours, its average speed is 30 mph (miles per hour). This speed is only an average as there would be times when the bus was travelling faster than this and times when it was travelling slower. There would even be times when it was not moving at all.

Key Point

Average speeds are worked out using the equation:

$$\text{average speed} = \frac{\text{distance travelled}}{\text{time taken}}$$

$$v = \frac{s}{t}$$

The symbol s is used for distance as d is used in other equations for diameter.
The symbol v is used for speed and velocity.

Here is an example of how to use this equation.

Calculate the average speed of a motorcycle that takes 6 seconds to travel 90 m.

Answer:

$$v = \frac{s}{t}$$

$$= \frac{90\text{ m}}{6\text{ s}}$$

$$= 15\text{ m/s}$$

Although we use miles and hours when talking about everyday journeys, the units used in science are metres (m) and seconds (s).

Progress Check

1 Complete the sentence:

To work out an average speed, measurements are needed of the distance travelled and the _______ _______.

2 Here are the times taken for three runners to complete a 100 m race.

Runner	Time in s
Mel	14.5
Sarfraz	15.4
Sam	12.5

Who won the race?

3 Calculate Sam's average speed.

1. time taken 2. Sam 3. 8.0 m/s

How long does it take?

In advance of a journey, you can only estimate the likely average speed.

If you are walking to catch a bus or travelling to an airport or ferry port, you need to know what time to set off so that you arrive in time. To estimate the time for a journey, two pieces of information are needed:

- the distance to travel
- the average speed

If you know how to re-arrange equations, you need only remember the speed equation in one form.

In good weather conditions, provided there are no delays such as traffic jams, the average speed of a car travelling on a motorway is about 60 mph. How long would a journey of 150 miles take?

To answer this you need to be able to use the speed equation in the form 'time = …'.

Key Point

The speed equation can be written in three different ways. In symbols, these are:

$v = \frac{s}{t}$ $t = \frac{s}{v}$ $s = v \times t$

The box on page 145 shows three different ways of writing the same equation. You can now work out the journey time using $t = \frac{s}{v}$.

You should get $2\frac{1}{2}$ hours.

Progress Check

Work out the quantities that go in the blank spaces in the table.

Take care to write the correct unit with your answer.

	Distance travelled	Time taken	Average speed
(a)	300 m	6 s	
(b)	6 cm	1.5 s	
(c)		5 hours	125 mph
(d)	1750 miles		500 mph
(e)	7.5 m	0.5 s	
(f)	1500 m		25 m/s
(g)		4.5 s	8 m/s

(a) 50 m/s (b) 4 cm/s (c) 625 miles (d) 3.5 hours (e) 15 m/s (f) 60 s (g) 36 m

Using graphs

If you walk to the end of the street and then back again, the distance that you have travelled is always increasing.

Two types of line graph that are often used to show an object's motion are:

- **distance-time** graphs – the distance travelled by an object can only increase with time
- **speed-time** graphs – speed can increase and decrease so the line can go down as well as up

Here are two graphs that show the same journey.

When answering questions about graphs that represent motion, always double check whether the graph is a distance-time graph or a speed-time graph.

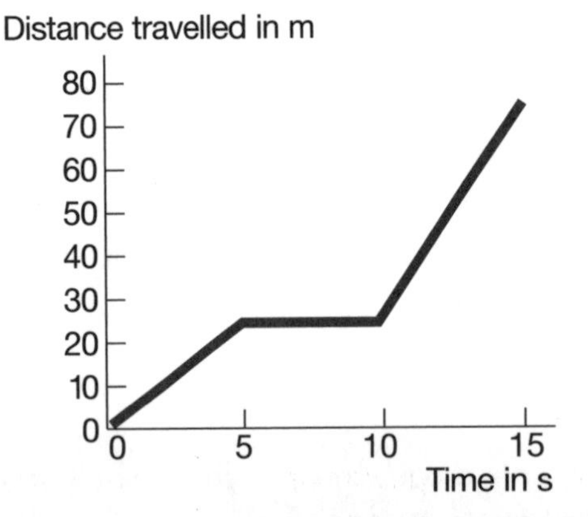

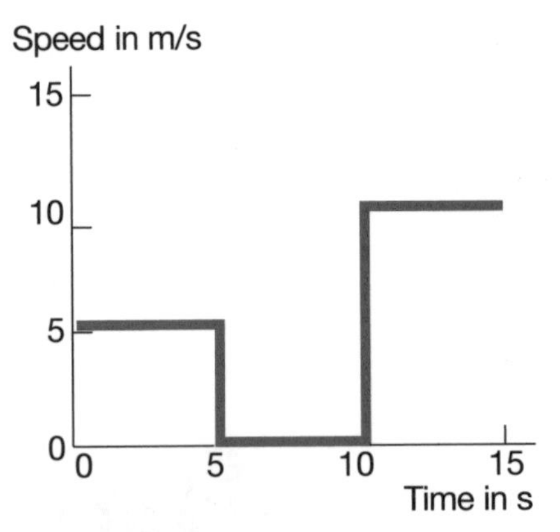

Fig. 9.13

If the graph line is horizontal then the object is not moving.

Both graphs show a constant speed of 5 m/s for the first 5 s, followed by a 5 s rest and then a constant speed of 10 m/s for the next 5 s. Notice how, on the distance-time graph, the steeper the line the greater the speed.

Key Point

The **slope**, or **gradient**, of a distance-time graph represents speed.

The greater the slope, the faster the object is moving.

Progress Check

Use words from the list to fill in the blanks.

decrease distance increase speed

Motion can be represented by a ________-time or a speed-time graph. The distance an object travels cannot _______, but its speed can ________or decrease.

The slope of a distance-time graph represents ________.

distance; decrease; increase; speed

9.4 Pressure and moments

After studying this topic you should be able to:

- **explain how the pressure caused by a force depends on the area that it acts on**
- **use the relationship between pressure, force and area**
- **describe the turning effect of a force and calculate its value**
- **apply the principle of moments**

Under pressure

Pushing a drawing pin into a board, cutting food with a knife and ice-skating are just three everyday examples of using a force to create a large pressure.

Pressure describes the effect a force has in cutting or piercing the surface it acts on:

- Knives, scissors, needles and drawing pins are all designed to cut or pierce. They create a large pressure by applying a force onto a small area.

Fig. 9.14

A common error in tests is to state that 'the pressure acts over a large area'. Do not confuse the terms 'force' and 'pressure'.

- Skis and caterpillar tracks on heavy vehicles are examples of spreading a force over a large area to reduce the pressure it causes.

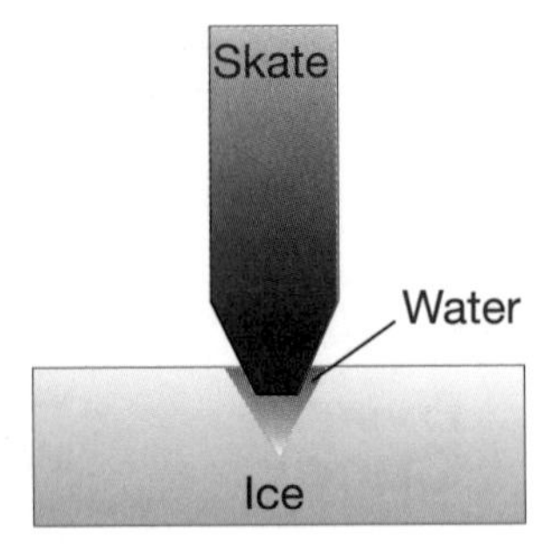

Fig. 9.15

Ice-skaters do not skate on ice; they skate on water.

- An ice-skater's weight pushes down on the small area of the blades.
- The high pressure on the ice below the blades causes it to melt, so the blades are surrounded by a film of water.
- The water re-freezes when the ice-skate has passed. This is how an ice-skater leaves tracks in the ice; the tracks are where the ice has melted and then re-frozen.

Progress Check

1 Complete the sentence:

To cause a high pressure, a force should act over a _______ area.

2 A drawing pin has a sharp point to cause a large pressure. True or false?

3 Explain the following:

(a) It is easier to cut with a sharp knife than with a blunt one.

(b) People are not allowed on bowling greens unless they are wearing flat-heeled shoes.

(c) It is more comfortable to sit on a soft cushion than on a hard stool.

1. small 2. True 3. (a) The sharp knife has a smaller surface area so it exerts a bigger pressure. (b) High-heeled shoes have a small surface area. The pressure they exert creates dents in the surface of the bowling green. (c) The cushion changes shape so there is more surface area in contact with the body, causing less pressure.

Quantifying pressure

Pressure is calculated as the force acting on each cm^2 or m^2 of surface area using the formula:

pressure = force ÷ area or $P = \frac{F}{A}$

Always give the correct unit, Pa or N/m^2

It is measured in N/m^2 or pascals (Pa). When the area involved is small, the unit N/cm^2 is used.

Here is an example:

A bulldozer weighs 150 000 N. To stop it from sinking into the soft mud it moves on caterpillar tracks. The area of the tracks in contact with the ground is 10 m^2. Calculate the pressure on the ground.

You need to be able to use all three forms of the equation to answer questions aimed at the highest levels.

Answer:

Pressure = force ÷ area
= 150 000 N ÷ 10 m^2
= 15 000 Pa

Like the speed equation, the pressure equation can be written in three different ways:

$$P = \frac{F}{A} \qquad F = P \times A \qquad A = \frac{F}{P}$$

Progress Check

Use the pressure equation to fill in the blanks in the table. Take care to write down the correct unit in each case.

	Force	Area	Pressure
(a)	25 N	2 m^2	
(b)		0.5 m^2	500 Pa
(c)	50 N		500 Pa
(d)	100 N	0.1 cm^2	
(e)		3 m^2	100 000 Pa

(a) 12.5 N/m^2 (b) 250 N (c) 0.1 m^2 (d) 1000 N/cm^2 (e) 300 000 N

Turning forces

Whenever we turn on a tap, push the pedals on a bike or open a door we are using a force to turn something round.

Key Point

The point that things turn around is called the pivot.

The pivot is the axis of rotation.

In the case of a tap, the pivot is at the centre of the tap. A door pivots around the hinge. The diagram shows the force and the pivot when a pedal is turned.

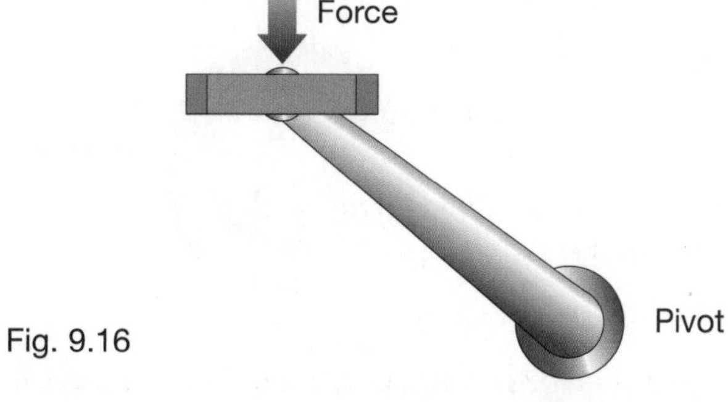

Fig. 9.16

If you try closing a door by pushing it at different places, you realise why door handles are placed as far from the pivot as possible.

The effect that a force has in turning something round is called the **moment** of the force. It depends on:

- the size of the force
- how far away from the pivot it is applied

Key Point

The further away from the pivot it is applied, the bigger the turning effect of a force.

The moment of a force is calculated using the relationship:

The perpendicular distance is the shortest distance from the force line to the pivot.

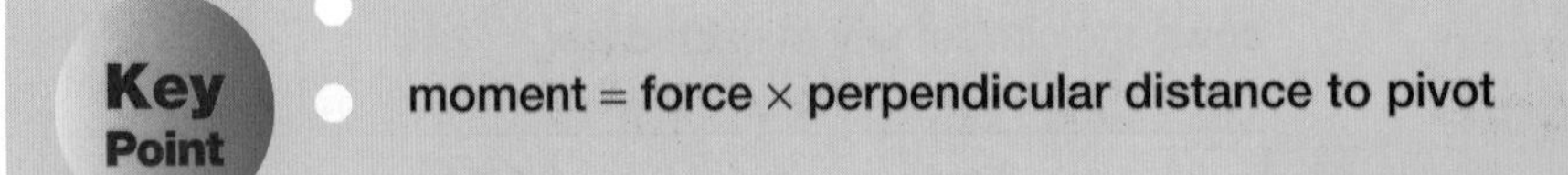

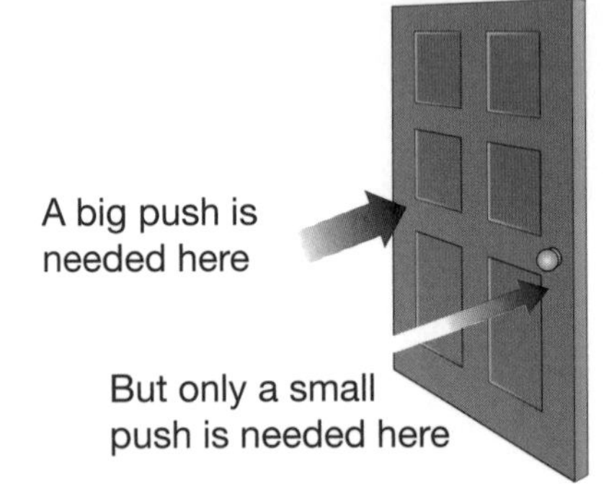

Fig. 9.17 The door handle is placed as far from the pivot as possible to increase the turning effect of the force.

1 Each of the diagrams below shows something that can be made to turn round. On each diagram, mark the pivot and a force that would cause turning.

2 Calculate the moment of the force being used to tighten the wheelnut.

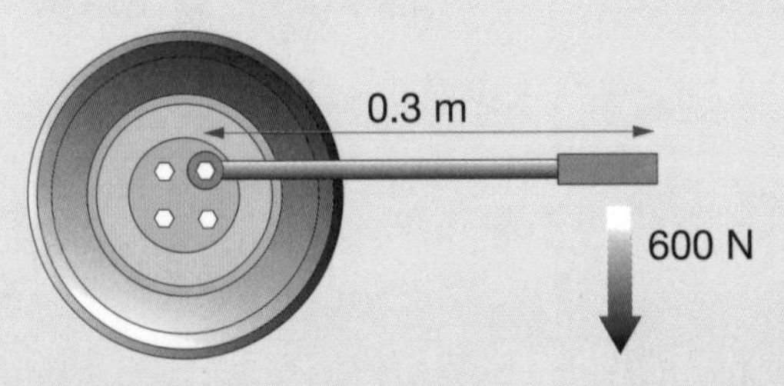

1. Door handle: the pivot is at the centre of the handle and the force should push up or down on the handle. Wheelbarrow: the pivot is at the base of the wheel and the force should push up on the handle. Spanner: the pivot is at the centre of the nut and the force should be to the left or the right. 2. 180 Nm

A question of balance

A seesaw has two forces acting on it:

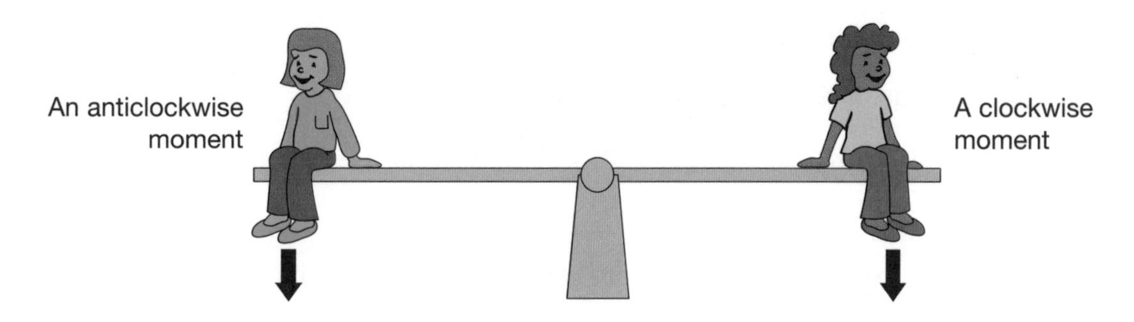

Fig. 9.18

Questions in tests will normally involve just one force on each side of the pivot.

- Each force has a turning effect but they are acting in opposition.
- If the moments of the forces are unequal, then the one with the bigger moment wins.
- If the forces have equal moments the seesaw is balanced.

This is an example of the **principle of moments**.

Key Point

The principle of moments states that:
If an object is balanced, the sum of the clockwise moments about a pivot is equal to the sum of the anticlockwise moments about the same pivot.

Progress Check

The diagrams show some seesaws. For diagrams A, B, C and D, decide whether each one is balanced, or whether it will rotate clockwise or anticlockwise. For diagrams E and F work out the size of the force needed to balance the seesaw.

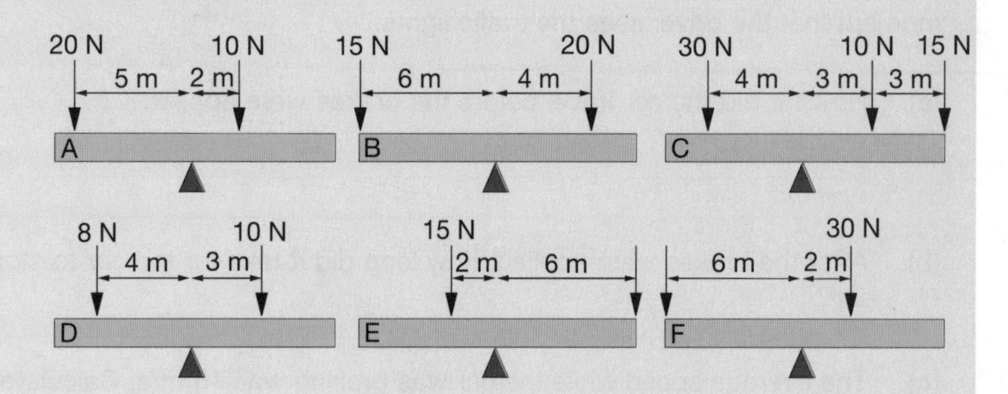

A: rotate anticlockwise; B: rotate anticlockwise; C: balanced; D: rotate anticlockwise; E: 5 N; F: 10 N

Practice test questions

The following questions test levels 3-6

A ball is thrown upwards. It rises and then falls down again.

(a) What happens to the kinetic energy of the ball as it rises? **[1]**

....

(b) At which point is the gravitational potential energy of the ball at its greatest? **[1]**

....

(c) Describe the energy transfer that takes place as the ball falls back to Earth. **[2]**

....

(d) The ball bounces several times and then stops.

What has happened to the ball's kinetic energy? **[1]**

....

(e) Underline two words in the list that describe the force between the Earth and the ball as the ball rises in the air. **[2]**

attractive electrostatic gravitational frictional magnetic repulsive

The following questions test levels 5-7

A car is travelling along a road at 30 m/s. The car driver sees some traffic lights at red and brakes to a halt. The graph shows how the speed of the car changes from the moment that the driver sees the traffic lights.

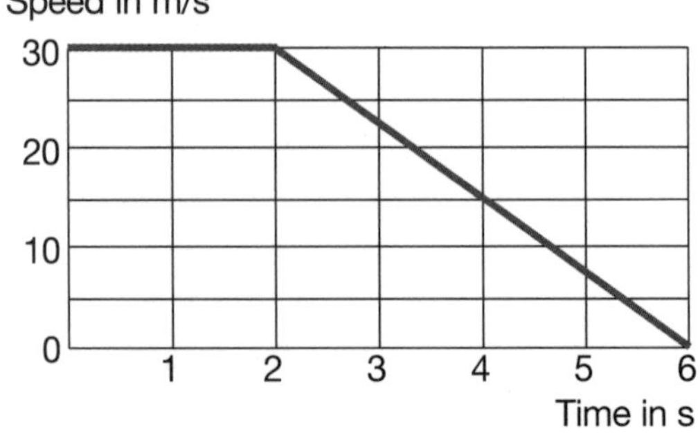

(a) How far did the car travel before the brakes were applied? **[3]**

....

....

(b) After the brakes were applied, how long did it take for the car to stop? **[1]**

....

(c) The average speed while the car was braking was 15 m/s. Calculate the distance the car travelled after the brakes were applied. **[2]**

....

....

(d) What was the total stopping distance of the car? **[1]**

....

Practice test answers

Chapter One

Levels 3–6

(a) (i) Plant plankton. [1]
Tip *This question is very common in National Tests. Remember, the producer in a food chain or web is the plant.*
(ii) Four. [1]
(iii) One. [1]
(iv) The numbers all decrease, [1]
as their food supply has been destroyed. [1]
Tip *Questions in National Tests often ask about how the changing numbers of one species affect the other species. An increased food supply leads to an increase in numbers of organisms further up the food chain.*

(b) (i) The fin allows the fish to propel itself through the water. [1]
(ii) The gill allows gas exchange. [1]
(iii) The stickles deter predators. [1]
Tip *You are not meant to have prior knowledge of a stickleback or any other organism. This question is testing whether you can apply your understanding of life processes to a new situation.*

Levels 5–7

(a) The cell has chloroplasts [1]
and a cell wall. [1]
Tip *These are the key differences between animal and plant cells. Remember, not all plant cells have chloroplasts – only those in the green parts of the plant. This is tested in **(c)**.*

(b) (i) The nucleus. [1]
(ii) The cell wall. [1]
(iii) The chloroplasts. [1]

(c) (i) The root. [1]
(ii) It has a large surface area to absorb water from the soil. [1]
Tip *This question is testing whether you can identify ways in which cells are adapted to do their job.*

Chapter Two

Levels 3–6

(a) Stomach C; lung A; small intestine E; large intestine D [4]

(b) Stomach. To break down food (digestion). [1]
Lung. Exchange of gases, oxygen and carbon dioxide, between lungs and bloodstream. [1]
Small intestine. Food chemicals are absorbed into the bloodstream. [1]
Large intestine. Water is taken out of the food and re-used. [1]

Levels 4–7

(a) Brand X [1]
Lower level of fat [1]
Lower level of carbohydrate [1]

(b) Sugar [1]

(c) These tests just show the presence of protein, carbohydrate and fat. They are all present in both yoghurts. It does not give amounts. [1]

(d) Protein 5.0 g; fat 3.13 g; carbohydrate 23.0 g [3]
Tip *You should show how you worked out your answers.*

Chapter Three

Levels 3–6

(a) (i) Smoking more cigarettes increases the risk of death from lung cancer. [1]
(ii) From the graph, 10 cigarettes a day increases the risk of death from lung cancer about 10 times [1]
(iii) Reducing smoking reduces the risk of lung cancer [1]

(b) Cigarette smoke contains tar [1]
Tar forms a layer in alveoli, reducing diffusion across cells [1]
Damages alveoli [1]

Levels 4–7

(a) Advantages
New plants will all be identical in colour and form; quicker than growing new plants from seed; cheaper. Any 2 [2]
Disadvantages
Any diseases in old plants present in the new. No chance of producing new colours. [2]
Tip *You must give two advantages and two disadvantages.*

(b) (i) There are no chloroplasts in the yellow parts of the leaf. [1]
Chlorophyll in chloroplasts needed for photosynthesis. [1]
(ii) carbon dioxide + water → glucose (or starch or sugar) + oxygen [2]
One mark for left-hand side and one mark for right-hand side.

Chapter Four

Levels 3–6

(a) D, C, A, E, B [3]
Tip *One mark if C anywhere before A, A anywhere before E and E anywhere before B.*
When you have finished, read the statements through again to check they are in the correct order.

(b) Ethanol is flammable so no naked flames should be present. [1]

(c) (i) Pink [1]
(ii) Pink [1]
(iii) Green [1]

(d) It is pink in both acid and neutral solutions. [1]

Levels 4–7

(a) Carbon dioxide [1]

(b) (i) Copper(**II**) carbonate + hydrochloric acid → copper chloride + water + carbon dioxide [2]
Tip *Writing word equations is at Level 6. There are two marks for three correct products and one mark for two.*
(ii) The mass of the products is less than the mass of the reactants. [1]
Because one of the products, carbon dioxide, has been lost as gas. [1]

(c) 8.0 g [1]
Tip *It does not matter which way you go from copper(II) carbonate to copper(II) oxide. The mass produced from a given mass of copper(II) carbonate is always the same.*

Chapter Five

Levels 3–6

(a) Silver colour [1]
Conductor of electricity [1]

(b) Paraffin wax melts over a range of temperatures; [1]
pure substances melt at a sharp temperature. [1]

(c) (i) Compound [1]
(ii) mercury oxide [1]

(d) (i) sulphur + oxygen → sulphur dioxide [2]
Tip *One mark for left-hand side and one mark for right-hand side.*
(ii) 3 [1]

Levels 4–7

(a) They both contain crystals. **[1]**
(b) It contains two types of crystals, white and black mixed together. **[1]**
(c) Rock A forms large crystals. These are formed when the magma crystallises slowly. **[1]**
(d) High pressure and high temperature. **[2]**
Tip *Two conditions are needed for two marks. You are not expected to give actual temperatures and pressures.*

Chapter Six

Levels 3–6

(a) (i) Carbon dioxide **[1]**
(ii) Dilute sulphuric acid **[1]**
(iii) All the acid is used up. **[1]**
(iv) Filtering **[1]**
(b) (i) There is a decrease in mass when zinc carbonate is heated. **[1]**
No change in mass when zinc oxide is heated. **[1]**
(ii) zinc oxide + carbon dioxide. One mark for each. **[2]**
Tip *The information in the table should suggest that when zinc carbonate is heated zinc oxide is formed.*

Levels 4–7

(a) To provide oxygen to start the reaction. **[1]**
Tip *Potassium chlorate splits up into potassium chloride and oxygen.*
(b) Friction provides the heat energy to start the reaction. **[1]**
(c) $S + O_2 \rightarrow SO_2$
$C + O_2 \rightarrow CO_2$ **[2]**
(d) Carbon dioxide and water **[2]**

Chapter Seven

Levels 3–6

(a) Any two from: oil, gas, coal. One mark each. **[2]**
Tip *Remember, a non-renewable source will run out within the lifetime of the Earth and cannot be replaced.*
(b) Any two from: wind, waves, tides, geothermal, solar. One mark each. **[2]**
(c) (i) On, off, on. One mark each. **[3]**
(ii) Move switch A from up to down. **[1]**
Move switch B from down to up. **[1]**
(iii) For switching a light on or off at either of two switches. **[1]**
Tip *This type of circuit is called a two-way switching circuit. It is commonly used for switching a landing light either upstairs or downstairs.*

Levels 5–7

(a) M is Mercury. **[1]**
V is Venus. **[1]**
(b) The Sun. **[1]**
Tip *Suns are stars that give out light because they are very hot. Planets and moons are seen by the light that they reflect.*
(c) Venus is close to the Sun, so it is illuminated brightly. **[1]**
Venus is also close to the Earth, so the light it reflects has spread out less than that of other planets when it reaches the Earth. **[1]**
(d) The side of the Earth facing away from the Sun should be shaded. **[1]**
(e) The Earth spins on its axis. **[1]**
Tip *The Earth rotates on its axis once each day. Take care not to confuse this with the Earth's movement around the Sun. One complete orbit around the Sun takes one year.*

Chapter Eight

Levels 3–6

(a) Heat passes through the steel by conduction **[1]**
as steel is a good conductor. **[1]**
Tip *When answering questions about thermal conduction, always stress whether the material is a good or a poor conductor. All metals are good conductors and all non-metals are poor conductors.*
(b) (i) By convection. **[1]**
(ii) The air next to the radiator is heated and expands. **[1]**
It rises as it is less dense than the surrounding air. **[1]**
Tip *This is an example of an upwards-driven convection current. Convection currents can also be driven by cold air moving down – for example in a fridge or next to a cold window.*
(c) Infra-red radiation. **[1]**

Levels 5–7

(a) The completed table is:

Colour in blue light	Colour in red light
blue	red
black	red
black	red

One mark for each correct answer. **[6]**
Tip *White objects always look to be the colour of the light shining on them. Objects of other colours act like filters; they absorb or subtract colours from the light. Anything that is not absorbed is reflected. If an object absorbs all the colours from a light, it appears black.*
(b) (i) The diagram should show the light reflected at the same angle as it hits. **[1]**
(ii) Tara's top scatters the light (reflects it in all directions). **[1]**
The mirror only reflects the light in the same direction as it hits it. **[1]**
Tip *This is a very common type of question in national tests. Take care to use the correct scientific term, 'reflect' and not 'deflect'.*

Chapter Nine

Levels 3–6

(a) It decreases or is transferred to gravitational potential energy. **[1]**
Tip *Kinetic energy is energy due to movement. Gravitational potential energy is energy due to position above the surface of the Earth.*
(b) At the greatest height. **[1]**
(c) Gravitational potential energy **[1]**
is transferred to kinetic energy. **[1]**
Tip *The energy transfer when the ball is falling is opposite to that when it is rising.*
(d) It is transferred to heat in the ball, the ground and the air. **[1]**
(e) Attractive **[1]**
gravitational **[1]**
Tip *All planetary movement is caused by the gravitational forces between very massive bodies. Gravitational forces only attract; they do not repel.*

Levels 5–7

(a) distance = speed x time **[1]**
= 30 m/s x 2.0 s **[1]**
= 60 m **[1]**
Tip *You need to have the correct unit to gain the third mark.*
(b) 4.0 s **[1]**
Tip *This is testing your skills at reading data from a graph. A common error is to answer '6 s', ignoring the 2.0 s before the car started braking.*
(c) distance = speed x time = 15 m/s x 4.0 s **[1]**
= 60 m **[1]**
(d) 120 m **[1]**
Tip *The total stopping distance is the 'thinking distance' plus the 'braking distance'.*

Investigating scientific questions

You need to understand how scientists in the past have asked questions and devised experiments to test the possible answers. You also need to understand how scientists work today and how they gather evidence to test predictions and explanations.

The past and the present

Most of the great scientific discoveries in history are well documented. Newton's theory of gravitation, Lavoisier's work on burning and Pasteur's discoveries are famous examples. In each case, their ideas could be tested to see if they were true or false.

The questions that face Science today are less easy to test, for example:

- Does the use of mobile telephones cause brain damage in young people?
- Can diseases such as BSE be passed from one species to another?
- Is human activity causing global warming?

Your investigative skills

As part of your work in Science you will have to carry out scientific investigations. These may be referred to as 'AT1' by your teacher. AT1 is the attainment target that sets out how you should progress in your investigative skills.

Your investigations will be marked by your teacher, who will assess your skills in:

- planning your investigation using your knowledge and understanding
- obtaining and presenting evidence
- considering evidence in reaching a conclusion
- evaluating your investigation

The results of your investigations will not be sent to an examiner outside the school at the end of Key Stage 3. An overall level will, however, be reported to your parents at the end of Key Stage 3 and will form the basis of similar work you will be required to do at Key Stage 4 for GCSE coursework.

Your scientific investigations do not require you to make new discoveries but to use your scientific knowledge, along with knowledge you can obtain from books, CD-ROMs and the Internet, to make a prediction. This prediction is then tested by experiments, which you have to plan and carry out. These experiments will produce results that you must record. Finally, you need to look at your results and see if they fit your original prediction.

After this you should evaluate your results and suggest any limitations they have, and any improvements that could be made to your experiment that would produce more reliable results.

What you need to do

1 Making predictions and planning your experiments

- Be sure you understand the scientific basis of any investigation you are carrying out.
- Ensure you make a clear prediction and give a reason why, based on your scientific knowledge and understanding. This could take the form of:
  I think that __________________ because ______________________________.
- Make a clear plan of what you are going to do and the equipment that you will need.
- List the observations and measurements that you are going to make.
- Plan a table to record your observations and measurements.
- Make sure that you have carried out a fair test. For example, if you are carrying out an investigation to compare the temperature rise of different samples of water with different fuels, have you planned to use the same mass (or volume) of water in each case?

2 Obtaining the evidence

- Carry out your experiments safely.
- Make all of the observations and measurements carefully.
- Repeat any observation or measurement if you have any doubt about it.
- Consider carrying out experiments more than once and averaging measurements.
- Record the observations or measurements as you carry out the experiment.

3 Looking at the results, drawing conclusions and considering the strength of your evidence

- Arrange your results in a suitable order and look for patterns.
- Draw suitable charts or graphs to display your results.
- Use the results to draw a conclusion.
- Use your scientific knowledge to try to explain the conclusion.

4 Evaluating your experiment

- Look critically at the conclusions to see if they are fully supported by the results.
- Identify any results that do not seem to fit the pattern.
- Suggest improvements that you would make if you carried out the experiment again.
- Explain how these improvements could improve the reliability of your conclusion.

The keys to success in scientific investigation are in making a sound prediction and carrying out a thorough evaluation. These are the aspects of your investigation that will be of particular interest to your teacher.

Most students enjoy this aspect of Science as they can use their scientific knowledge and understanding to make and test predictions in the same way a scientist does.

Science word list

For success in KS3 tests you must use the language of Science correctly. This list includes the words you should know and use. As part of your revision, look in this book to check that you understand how to use each word and spell it correctly.

Biology

absorption
adolescence
alveolus (plural alveoli)
amphibian
antagonistic muscle pair
antibiotic
anus
asexual reproduction
bacterium (plural bacteria)
Benedict's solution
biuret
bronchiole
bronchus
capillary
carbohydrate
carnivore
cell
cell membrane
chlorophyll
chloroplast
chromosome
cilia
classification
cuticle
diaphragm
diet
duodenum
embryo
environment
environmental factor
epidermis
fat
fertilisation
fetus
fibre
food chain
food web
gamete
gene
genetic information
habitat
haemoglobin
herbivore
immune
intercostals muscle
invertebrate
large intestine
larynx
liver
lung
mammal
meiosis
menstrual cycle
microbe
micro-organism
mineral
mitosis
mutation
nucleus (plural nuclei)
oesophagus
omnivore
organ
ovary
ovum (plural ova)
palisade cell
pancreas
penis
photosynthesis
placenta
predator
pregnancy
prey
producer
protein
rectum
reptile
respiration
rib
scrotum
selective breeding
sex cells
sexual reproduction
skeleton
small intestine
species
sternum
stomach
stomata (plural stoma)
synovial joint
testes
tissue
toxin
trace element
trachea
umbilical cord
uterus
vacuole
vagina
variation
vertebrate
virus
vitamin
vulva
zygote

Chemistry

acid
alkali
alloy
antacid
atom
brittle
burning
chemical change
chromatography

combine
combustion
compound
conglomerate
crystal
crystallisation
diffusion
distillation
ductile
element
endothermic
erosion
evaporation
exothermic
formula
fossil
fuel
gas
global warming
grain
greenhouse effect
hazard sign
igneous
indicator
insoluble
liquid
litmus
magma
malleable
metal
metamorphic
mineral
mixture
molecule
neutral
neutralisation
non-metal
periodic table
permeable
pH value
pollution
porosity
reactivity series
rock cycle
rusting
salt
saturated solution
sedimentary
shale
simple cell
solid
soluble
solution
symbol
Universal Indicator
weathering

Physics

air resistance
ammeter
amplitude
armature
asteroid belt
axis
balanced forces
circuit
coil
compression
conduction
conductor
convection
core
current
dispersion
driving force
eardrum
electricity
electromagnet
electromagnetic
ellipse
energy
force
fossil fuel
frequency
friction
generator
geothermal energy
gravitational force
infra-red
insulation
insulator
joule
kinetic energy
loudspeaker
luminous
magnet
magnetic field
mass
moment
newton
non-renewable
oscilloscope
ossicles
parallel
pitch
pivot
planet
pole
pressure
primary colours
prism
rarefaction
real image
reflection
refraction
relay
renewable
resistance
resistive force
satellite
scatter
secondary colours
series
solar cell
Solar System
sound
space
speed
star
temperature
turbine
unbalanced forces
variable resistor
velocity
vibrate
virtual image
voltage
weight

Index